BIOLOGICAL AND MEDICAL PHYSICS, BIOMEDICAL ENGINEERING

BIOLOGICAL AND MEDICAL PHYSICS, BIOMEDICAL ENGINEERING

The fields of biological and medical physics and biomedical engineering are broad, multidisciplinary and dynamic. They lie at the crossroads of frontier research in physics, biology, chemistry, and medicine. The Biological and Medical Physics, Biomedical Engineering Series is intended to be comprehensive, covering a broad range of topics important to the study of the physical, chemical and biological sciences. Its goal is to provide scientists and engineers with textbooks, monographs, and reference works to address the growing need for information.

Books in the series emphasize established and emergent areas of science including molecular, membrane, and mathematical biophysics; photosynthetic energy harvesting and conversion; information processing; physical principles of genetics; sensory communications; automata networks, neural networks, and cellular automata. Equally important will be coverage of applied aspects of biological and medical physics and biomedical engineering such as molecular electronic components and devices, biosensors, medicine, imaging, physical principles of renewable energy production, advanced prostheses, and environmental control and engineering.

More information about this series at http://www.springer.com/series/3740

Eugenijus Kaniusas

Biomedical Signals and Sensors II

Linking Acoustic and Optic Biosignals and Biomedical Sensors

With 73 Figures

Springer

Ao. Univ.-Prof. Dipl.-Ing. habil. Dr. Eugenijus Kaniusas
Head of research group 'Biomedical Sensors'
Vienna University of Technology
Institute of Electrodynamics, Microwave and Circuit Engineering
Gusshausstr. 27–29
1040 Vienna
Austria
E-mail: kaniusas@tuwien.ac.at

Volume 1: ISBN 978-3-642-24843-6

ISSN 1618-7210 ISSN 2197-5647 (electronic)
Biological and Medical Physics, Biomedical Engineering
ISBN 978-3-662-45105-2 ISBN 978-3-662-45106-9 (eBook)
DOI 10.1007/978-3-662-45106-9

Library of Congress Control Number: 2012930477

Springer Heidelberg New York Dordrecht London

Printed on acid-free paper

Springer-Verlag GmbH Berlin Heidelberg is part of Springer Science+Business Media
(www.springer.com)

Preface

The present volume set develops a bridge between physiologic mechanisms and diagnostic human engineering. A multitude of biomedical sensors are commonplace in clinical practice today. The registered biomedical signals, which will be referred to as biosignals, reflect vital physiologic phenomena and are relevant not only for diagnosis but also for therapy. In order to adequately apply biomedical sensors and reasonably interpret the corresponding biosignals, a proper and strategic understanding of the physiologic phenomena involved, their specific influence on the registered biosignals, and the technology behind the sensors is critical.

While the *first volume* is focused on the interface between physiologic mechanisms and the resultant biosignals, the second volume is devoted to the interface between biosignals and biomedical sensors. That is, in the first volume, the physiologic mechanisms determining biosignals are described from the basic cellular level—as the place of origin of each and every biosignal—up to their advanced mutual coordination level, e.g., during sleep. It allows a physiologically accurate interpretation and comprehensive analysis of biosignals.

Consequently, this *second volume* considers the genesis of acoustic and optic biosignals and the associated sensing technology from a *strategic point of view*. Unlike other contributions, this book deals differently on the subject of specific engineering aspects pertaining to particular biosignals, since it discusses *heterogeneous biosignals* within a *common frame*. In particular, this frame comprises both the *biosignal formation path* from the biosignal source at the physiological level to biosignal propagation in the body and the *biosignal sensing path* from the biosignal transmission in the sensor applied on the body up to its conversion to a, usually electric, signal.

Some biosignals arise in the course of the body's vital functions while others map these functions that convey physiological data to an observer. It is highly instructive how sound and light beams interact with biological tissues, yielding *acoustic and optic biosignals*, respectively. Discussed phenomena teach a lot about the physics of sound and physics of light (as engineering sciences), and, on the other hand, biology and physiology (as live sciences). This book provides a sort of *common denominator* for acoustic and optic biosignals, i.e., instructive similarities and differences in between, whereas these biosignals—at first glance—are entirely different in their physical nature. Basic and application-related issues are covered in depth; in fact, these issues should remain strong because these stand the test of time and mine knowledge of great value.

This book is directed primarily at graduate and post graduate students in biomedical engineering and biophysics. It is also accessible to those who are interested in physical, engineering, and life sciences, since expected background knowledge is minimal and many basic phenomena are explained in depth within numerous footnotes. Furthermore, the book should serve engineers and practitioners who have an interest in biomedical engineering. Discussed biosignals and sensing technologies substantiate wearable sensor technologies—the hot topic today—which comprise an appealing solution for pervasive (home) monitoring and prompt novel approaches in diagnosis and therapy.

It is important to note that this book was mainly inspired by my lectures entitled "Biomedical Sensors and Signals," "Biomedical Instrumentation," and "Biophysics" which constitute a significant part of a master's degree program "Biomedical Engineering" at the Vienna University of Technology in Austria.

In the end, it is not the spot-like knowledge of biosignals and engineering technologies coming from independent considerations of biosignals that constitute a successful biomedical engineer with profound professional knowledge, but the strategic and global consideration of basically different biosignals and of the corresponding sensing technologies, both integrated in the common frame. The highly interdisciplinary nature of biosignals and biomedical sensors is obviously a challenge. However, it is a rewarding challenge after it has been coped with in a strategic way, as offered here. The book is intended to have the presence to answer intriguing "Aha!" questions.

Vienna, Austria Eugenijus Kaniusas

Acknowledgments

I would like to express my sincerest thanks for support to Univ. Prof. Helmut Pfützner (from the Institute of Electrodynamics, Microwave and Circuit Engineering (EMCE), Vienna University of Technology, Austria), Univ. Prof. Giedrius Varoneckas (from the Sleep Medicine Centre at Klaipeda University Hospital, Lithuania), Univ. Prof. Bernd Saletu (from the Department of Psychiatry, University of Vienna, Austria), and Dr. Lars Mehnen, Dr. Karl Futschik, Dr. Stefan Traxler (all from the EMCE).

I thank my students Dipl.-Ing. Stefan Kampusch, Florian Thürk, and Dejan Tomic for experimental support and careful proof reading. The book has significantly benefited from countless small and large projects in which numerous diligent students of mine have been involved. I give sincere thanks to all of them. Special thanks go to Dipl.-Ing. Daniel Von-Chamier-Glisczinski who prepared ball pen drawings prefacing each chapter and to Univ. Prof. Andrius Baltuska for scientific advice. I express my deep gratitude to my family, parents, and other relatives for unlimited support.

Contents

Symbols and Abbreviations

Note: Variables used within limited contexts are not listed, for they are described within the relevant section. The different types of biosignals are separately listed below.

A	Area
BMI	Body mass index
c	Constant, specific heat capacity, speed of light (in vacuum)
C	Capacitance
CSA	Central sleep apnea
d	Distance, light probing depth
D	Compliance, diffusion length
E	Electric field
E_{I}	Incident electric field
E_{R}	Reflected electric field
f	(oscillating, resonating) Frequency
f_{C}	Heart rate
f_{F}	Formant frequency
f_{R}	Respiratory rate
f_{R1}	Fundamental harmonic frequency of lung sounds or snoring sounds
g	Scattering anisotropy coefficient
G	Transfer function
h	Planck's constant
H	Hematocrit
I	Electric current amplitude, sound intensity, light intensity
I_{AC}	Alternating component of light intensity
I_{DC}	Direct component of light intensity
k	Index, wavenumber
l	Length
LED	Light-emitting diode

MSA	Mixed sleep apnea
n	Index of refraction
OHA	Obstructive sleep hypopnea
OSA	Obstructive sleep apnea
p	Air pressure, sound pressure, power spectral density, probability density, blood pressure
P	Power, acoustic pulsatile pressure amplitude
p_I	Incident pressure wave
p_R	Reflected pressure wave
$p_\mathrm{S,D}$	Systolic–diastolic deflection of the blood pressure
p_T	Transmural pressure
q	Air flow
Q	Electric charge
r	Radius, distance
r_D	Diastolic artery radius
r_S	Systolic artery radius
r_R	Source-sink distance in the reflectance mode
r_T	Source-sink distance in the transmittance mode
R	Reynolds number, (electrical) resistance, red to infrared absorbance ratio
$\mathcal{R}$	Alternating to direct light ratio
s	Biosignal, see below
s^C	Cardiac component of biosignal
s^R	Respiratory component of biosignal
$s_\mathrm{S,D}$	Systolic–diastolic deflection of the cardiac component
S	Biosignal amplitude, hemoglobin oxygen saturation
SPL	Sound pressure level
t	Time
u	Voltage, air flow velocity, particle velocity
U	(complex) Voltage amplitude
v	Sound propagation velocity, light propagation velocity, pulse wave velocity
V	Volume
W	Energy
x	Coordinate, distance
X	Distance
Z	(complex) Electrical impedance, characteristic acoustic impedance
α	Sound absorption coefficient, light attenuation coefficient, constant
Γ	Reflection factor
ε	Dielectric permittivity, step function
ε_r	Relative electric permittivity
ϑ	Temperature
κ	Module of volume elasticity
λ	Wavelength
μ	Dynamic viscosity, magnetic permeability
μ_A	Light absorption coefficient

μ_r	Relative magnetic permeability
μ_S	Light scattering coefficient
μ_S'	Light reduced scattering coefficient
μ_T	Light total absorption coefficient
ρ	Density
σ	Mechanical stress, absorption cross section
τ	Relaxation time constant, time constant
υ	Heat conductivity
φ	Angle
ω	Angular frequency

Symbols of Biosignals

The types of biosignals discussed and their short descriptions.

Symbol	Name	Biosignal class		Phenomena reflected
s_{ECG}	electrocardiogram signal	permanent	electric	electrical excitation of heart muscles
s_{MRG}	mechanorespirogram signal		mechanic	circumference changes of the abdomen or chest during breathing
s_{PCG}	phonocardiogram signal		acoustic	sounds emitted by sources in the inner body
s_{OPG}	optoplethysmogram signal	induced	optic	artificial light absorption by pulsatile blood

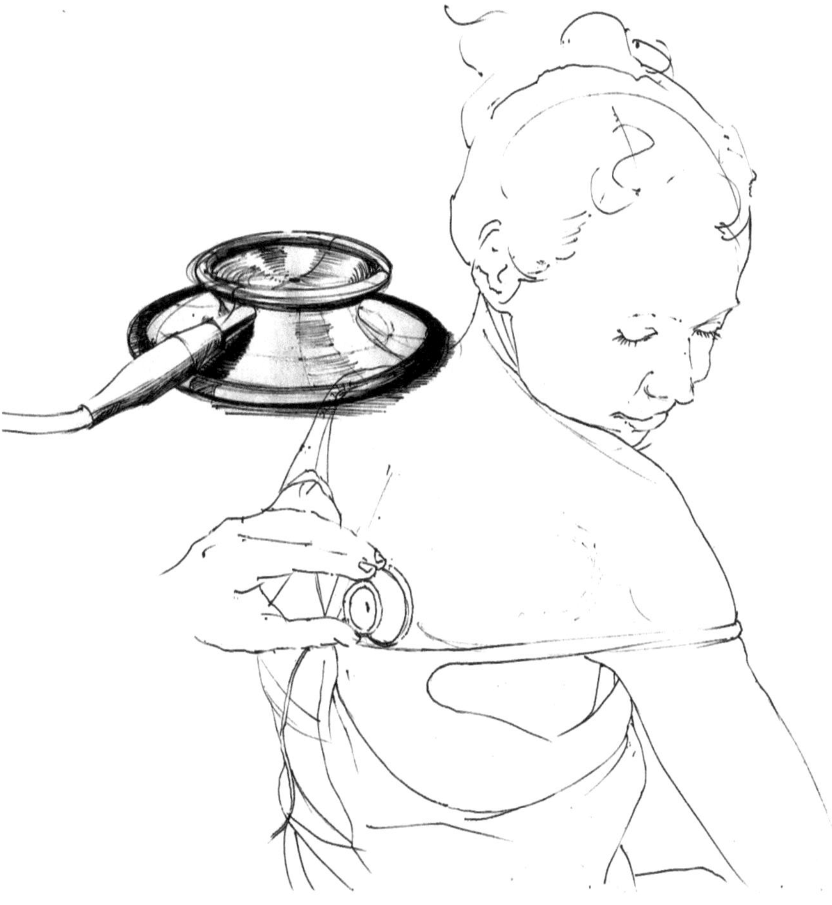

(Chamier, 2014)

Chapter 4
Sensing by Acoustic Biosignals

Abstract After the interface between physiologic mechanisms and the resultant biosignals has been examined (Volume I), the subsequent interface between acoustic biosignals and the associated sensing technology is discussed here. A large variety of acoustic biosignals—permanent biosignals—originates in the inner human body, including heart sounds, lung sounds, and snoring sounds. These biosignals arise in the course of the body's vital functions and convey physiological data to an observer, disclosing cardiorespiratory pathologies and the state of health. The genesis of acoustic biosignals is considered from a strategic point of view. In particular, the introduced common frame of hybrid biosignals comprises both the biosignal formation path from the biosignal source at the physiological level to biosignal propagation in the body, and the biosignal sensing path from the biosignal transmission in the sensor applied on the body up to its conversion to an electric signal. Namely, vibrating structures in the body yield acoustic sounds which are subject to damping while propagating through the thoracic tissues towards the skin. Arrived at the skin, different body sounds interfere with each other and induce mechanical skin vibration which, in turn, is perceived by a body sound sensor and then converted into the electric signal. It is highly instructive from an engineering and clinical point of view how sounds originate and interact with biological tissues. Discussed phenomena teach a lot about the physics of sound (as engineering sciences), and, on the other hand, biology and physiology (as live sciences). Basic and application-related issues are covered in depth. In fact, these issues should remain strong because these stand the test of time and mine knowledge of great value. Obviously, the highly interdisciplinary nature of acoustic biosignals and biomedical sensors is a challenge. However, it is a rewarding challenge after it has been coped with in a strategic way, as offered here. The chapter is intended to have the presence to answer intriguing "Aha!" questions.

E. Kaniusas, *Biomedical Signals and Sensors II*,
Biological and Medical Physics, Biomedical Engineering,
DOI 10.1007/978-3-662-45106-9_4

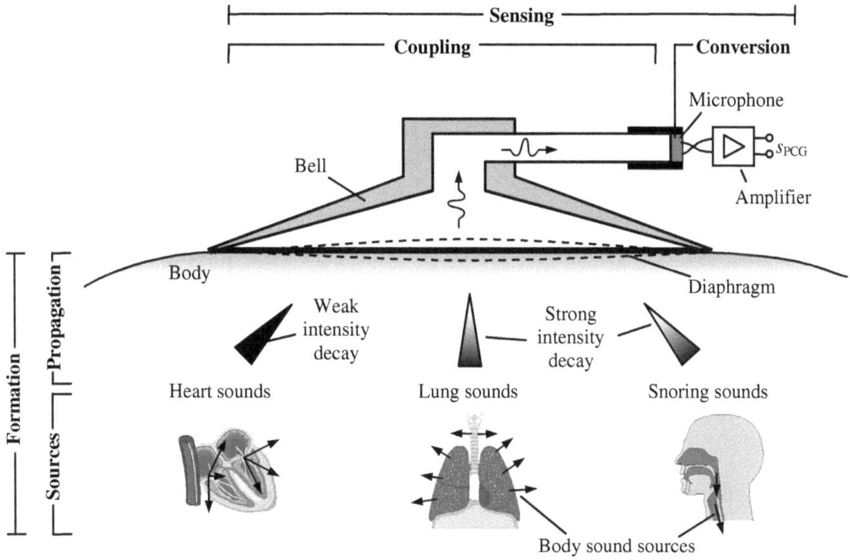

Fig. 4.1 The stethoscope chestpiece is applied at the chest for the auscultation of various body sounds. The formation and sensing path of an acoustic biosignal phonocardiogram s_{PCG} is depicted; compare Fig. 4.2. The formation path includes sources of body sounds and the damping of sounds during their propagation through the body tissues, whereas the sensing path includes coupling and conversion of sounds

A large variety of *acoustic biosignals*, i.e., *permanent biosignals* according to their classification (Sect. 1.3), originates in the inner human body, including heart sounds, lung sounds, and snoring sounds. The *auscultation* of these *body sounds* is a timeless classic for diagnosis of health status, especially since Dr. Laennec, the inventor of the stethoscope, fundamentally improved the auscultation technique (Sect. 1.2). The body sounds convey numerous meaningful signals to the physician, disclosing *cardiorespiratory* pathologies or the state of health; compare with multiparametric monitoring (Sect. 1.4).

Traditionally, the auscultation of *heart sounds* is applied to detect cardiac pathologies. Auscultation of *lung sounds* and snoring sounds is applied to detect respiratory disturbances. Recently, medical interest has also focused on *snoring sounds* as an important symptom of the sleep apnea syndrome, i.e., a temporal and repetitive cessation of effective respiration during sleep at night (Sect. 3.1.2).

Figure 4.1 demonstrates a *body sound sensor* on the skin, which is basically given by the *chestpiece of the stethoscope* combined with a microphone. *Vibrating structures* in the body yield acoustic sounds which are subject to damping while *propagating through the thoracic tissues* towards the skin. Arrived at the skin, different body sounds interfere with each other and induce mechanical skin vibration which, in turn, forces the chestpiece diaphragm to oscillate. This oscillation creates acoustic *pressure waves* travelling into the chestpiece bell and down

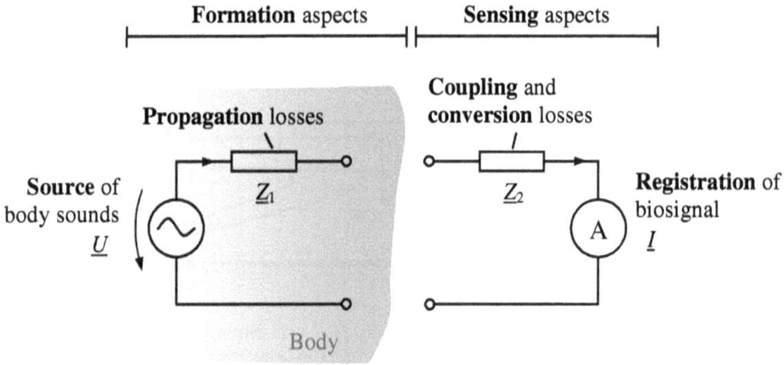

Fig. 4.2 Model of permanent acoustic biosignal, including its generation, propagation, coupling, and registration; compare Fig. 1.3

to the microphone. The microphone serves as an acousto-electric converter to establish an *acoustic biosignal* s_{PCG}.

The *formation of body sounds* up to their registration can be simplified as an *electrical circuit model*, as illustrated in Fig. 4.2 (compare Sect. 1.1). In accordance with this model and in analogy with Fig. 4.1, we start with *sources* of body sounds (represented by voltage source $\underline{U}$ in Fig. 4.2) and go over the *propagation* of generated sounds (propagation losses represented by electrical impedance $\underline{Z}_1$ in Fig. 4.2) throughout biological tissue. As a certain portion of sounds leaves the body and thus is available for its auscultation, we continue with the sound *coupling* (coupling represented by electrical impedance $\underline{Z}_2$ in Fig. 4.2) into an acoustic sensing device, i.e., a body sound sensor applied on the skin. Lastly, the sounds *conversion* (conversion losses as an additive part of electrical impedance $\underline{Z}_2$ in Fig. 4.2) into an electric signal is modelled, preceding the *registration* of acoustic biosignals (modelled as ampere meter in Fig. 4.2).

4.1 Formation Aspects

According to Figs. 4.1 and 4.2, *formation aspects* include

- the *genesis of body sounds* and
- their *transmission* in the body

towards the sensing device applied on the skin. The formation aspects reveal not only *clinically relevant correlations* between physiological phenomena of interest and recorded acoustic biosignals but also facilitate a proper understanding of the *biosignal's diagnostical relevance.*

In particular, distinctive types of *sources of body sounds* are reviewed, commenting on their mutual interrelations. It will be shown that body sounds, i.e., *mechanical waves* within the body, originate in the course of *mechanical vibrations*

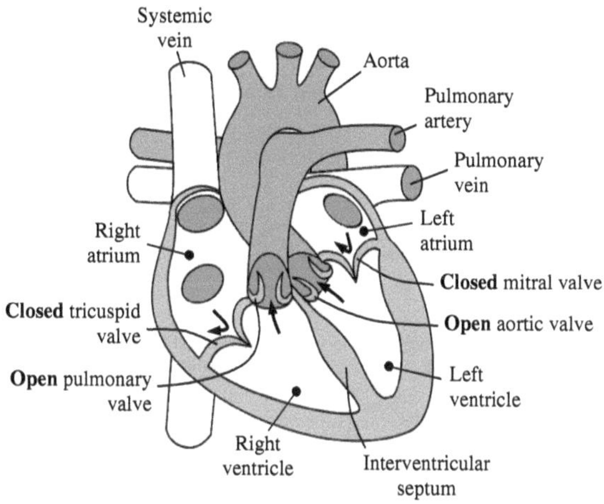

Fig. 4.3 Anatomic structure of the heart relevant for the generation of heart sounds; compare Fig. 2.32. Image data partly taken from Wikipedia (2010)

of *tissues and blood*, vibrations of *heart valves*, oscillation of *airway walls*, and air *turbulences* in the airways. From an acoustical point of view, body sounds comprise *impure tones or noises* and thus are composed of multiple spectral components of varying intensity and frequency. Later, the *transmission of body sounds* in the human body is discussed in the time and spatial domain, focussing on the sound's interaction with the biological medium.

4.1.1 Body Sounds—An Overview

4.1.1.1 Heart Sounds

Heart sounds are probably the most familiar body sounds which auscultation was fundamentally improved by the invention of the stethoscope (Sect. 1.2.1). In general, these sounds are related to the contractile activity of the cardiac system, including the heart and blood together,[1] and the *blood's turbulence* in atria and ventricles of the heart (Fig. 4.3). The sounds yield direct information on *myocardial contractility* and the *valve's closure* in the heart (Sect. 2.4.1).

In particular, heart sounds are generated within or close to the heart (Fig. 4.1). These sounds can be roughly classified into (Kaniusas 2007; Walker et al. 1990; Rangayyan 2002; Lessard and Jones 1988; Amit et al. 2009)

- normal heart sounds and
- abnormal heart sounds.

[1] The *heart and blood* together is comparable to a thin-walled and fluid-filled *balloon* which, when stimulated at any location, starts to vibrate as the whole and to *emit sounds* (Rangayyan 2002).

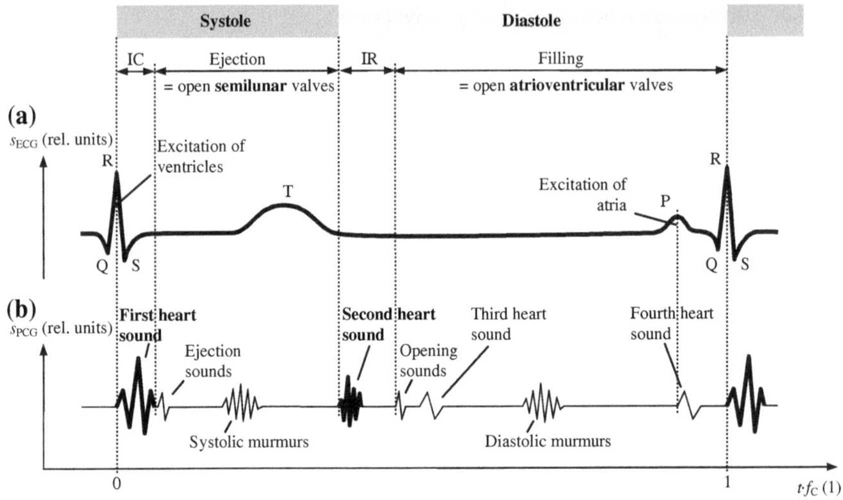

Fig. 4.4 Schematic waveforms of (**a**) electric biosignal electrocardiogram s_{ECG} with typical waves and peaks in relation to (**b**) acoustic biosignal phonocardiogram s_{PCG} during the cardiac cycle with the duration $1/f_C$; compare Fig. 2.38. Typical phases of the cardiac cycle are denoted including the isovolumetric contraction phase (IC) and the isovolumetric relaxation phase (IR). Different heart sounds are depicted with normal sounds drawn in *bold*, i.e., the first and second heart sounds. The frequency and amplitude of the respective heart sounds are qualitatively indicated

Within a single cardiac cycle, the following *normal heart sounds* can be observed, as illustrated in Fig. 4.4b (compare Fig. 2.38):

- The *first heart sound*: it is associated with the *closure of atrioventricular valves*, i.e., the tricuspid and mitral valves, preventing the backward flow of blood (from ventricles into atria). Abrupt tension changes of atrioventricular valves, deceleration of the blood flow, and jerky contraction of the ventricular myocardium induce mechanical vibrations (with reverberations) of the involved structures, which manifest as the first heart sound. This sound occurs at the onset of the ventricular *systole* immediately after the R wave of the electrocardiogram (Fig. 4.4a). The first sound is usually composed of *two components* (sound *splitting*), caused by the closure of the right-sided tricuspid valve and the left-sided mitral valve, respectively (Fig. 4.3). The mitral valve closes slightly earlier than the tricuspid valve,[2] so that the left-sided sounds slightly precede by about

[2] The *asynchronous closure of atrioventricular valves* can be attributed to several factors (Brooks et al. 1979). In particular, the left ventricle contracts slightly before the right ventricle, yielding an earlier closure of the mitral valve (Fig. 4.3). In addition, the mitral valve is more (nearly) closed when the contraction of the left ventricle begins than is the tricuspid valve when the contraction of the right ventricle begins; the asynchronous closure also greatly depends on the contraction and relaxation of both atria. The *inspiration* also delays the closure of the tricuspid valve because of increased venous return (see section "Normal Respiration" in Sect. 3.2.1.1), which enhances the splitting of the first heart sound (see section "Normal Respiration" in Sect. 3.2.1.2).

20 ms (Walker et al. 1990). Normally, the first heart sound is the *loudest and longest* of all heart sounds, including spectral components of *relatively low frequency* (Footnote 150 in Sect. 2). The sound's duration is about 140 ms with the spectral peak at about 30 Hz, whereas the intensity decreases by about 40 dB in the range of 10–100 Hz.

- The *second heart sound*: it is associated with the *closure of semilunar valves*, i.e., the pulmonary and aortic valves, preventing the backward flow of blood (from arteries into ventricles). In analogy with the first heart sound, the valve's closure and abrupt deceleration of the blood flow yield *vibrations of the valve's cusps* and blood in the great vessels. In particular, when the elastic limits of the tensed valve leaflets are met oscillations of the leaflets are initiated. Atria and ventricles vibrate in concert with the valves and blood because of their anatomical vicinity. This sound occurs at the onset of the ventricular *diastole*, beginning at the end of the T wave of the electrocardiogram (Fig. 4.4a). The second sound is usually composed of *two components* (sound *splitting*), caused by the closure of the right-sided pulmonary valve and the left-sided aortic valve, respectively (Fig. 4.3). Typically the aortic valve closes earlier than the pulmonary valve because of *respiration effects* (see section "Normal Respiration" in Sect. 3.2.1.2), so that the left-sided sounds precede by about 40 ms during *inspiration* (Walker et al. 1990); with *expiration* the left-sided and right-sided sounds may be superimposed or even still slightly split by < 30 ms. The left-sided sounds are usually louder[3] than the right-sided sounds due to a much higher blood pressure in the aorta at the onset of diastole. The second heart sound shows shorter duration of about 110 ms (< 140 ms), lower intensity, *higher frequency components*, and a more *snapping quality* than the first heart sound. Its short duration and dominant high frequency components result from the fact that semilunar valves are much tauter than atrioventricular valves and thus close more rapidly. In contrast to the first heart sound, the second sound does not show a consistent spectral peak and is not limited to a relatively narrow frequency bandwidth, whereas the intensity decreases more slowly by about 30 dB (< 40 dB) in the range of 10–100 Hz.

Normally, the *first and second heart sounds* are audible only. Figure 4.4 depicts that the time period between the first heart sound and the second heart sound defines the *duration of the ventricular systole*. Early studies found that the *spectral components* are negligible above 110 Hz in normal heart sounds (Rappaport and Sprague 1941) or reside in the approximate range of 20–120 Hz (Abella et al. 1992).

[3] The significantly higher pressures on the left side of the heart cause the *left-sided valves* to shut harder and faster than the closure of the *right-sided valves*. Therefore, the majority of auscultated heart sounds originates from the left-sided valves; though this can not be generalised and depends strongly on the auscultation location (compare section "Normal Respiration" in Sect. 3.2.1.2).

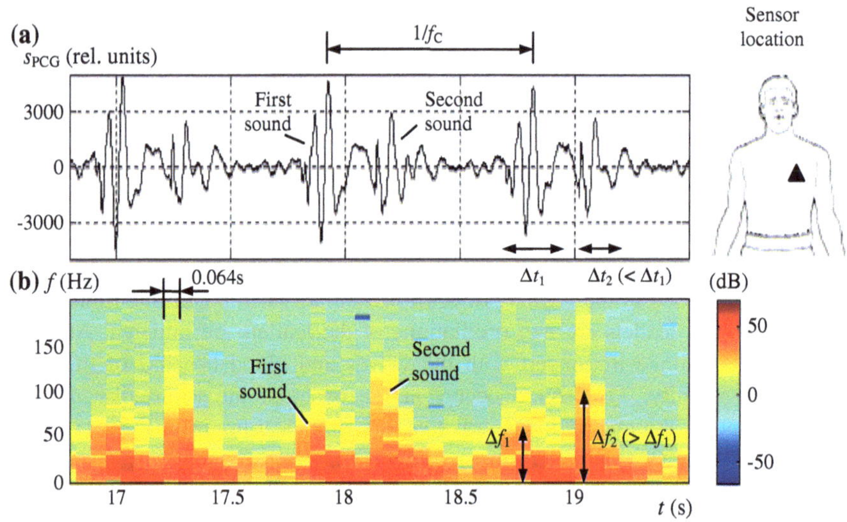

Fig. 4.5 Normal heart sounds while holding breath. (**a**) Acoustic biosignal phonocardiogram s_{PCG} from the heart region at the chest with indicated heart rate f_C. (**b**) The corresponding spectrogram (see Footnote 4) shows differences between the first and second heart sound ($\Delta t_1 > \Delta t_2$ and $\Delta f_1 < \Delta f_2$)

Figure 4.5 illustrates normal heart sounds while holding breath, which reoccur with the heart rate f_C; for *sensing aspects* of body sounds see Sect. 4.2. It can be observed in Fig. 4.5a that the first heart sound exhibits a larger signal deflection and is a bit longer in duration than the second heart sound ($\Delta t_1 > \Delta t_2$). The corresponding spectrogram[4] in Fig. 4.5b demonstrates that both *normal heart sounds* are characterized by short-term frequency components mainly in the frequency range of *up to about 100 Hz*, with weak contributions up to 400 Hz (Kaniusas et al. 2005); compare Fig. 4.8b. However, the second heart sound includes spectral components of higher frequency than the first heart sound ($\Delta f_1 < \Delta f_2$ in Fig. 4.5b). In addition, the *50 Hz interference* from power lines can be recognised. The discussed *splitting* of the second heart sound is illustrated in Fig. 3.32.

[4] A *spectrogram* provides information of how signal power (i.e., signal variance) is distributed as a function of *frequency and time*; compare Footnote 193 in Sect. 3. Likewise, the spectrogram shows how the *power spectral density* varies with time. As illustrated in Fig. 4.5b, the horizontal axis represents *time t* while the vertical axis *frequency f*. The dot color in the image (i.e., the third dimension) reflects the signal power in a *logarithmic scale* for a particular frequency and at a particular time instant, see the color bar to the right of Fig. 4.5b. The striped pattern of the spectrogram in the vertical direction results from the fact that the power spectral density was calculated for time intervals of 0.128 s duration with 50 % overlap, yielding a time resolution of 0.064 s (Fig. 4.5b) and a frequency resolution of about 1/0.128 s ≈ 7.8 Hz.

Numerous *other heart sounds* originate within the scope of cardiac activity, which prevalence and audibility depend on age or indicate abnormality. As illustrated in Fig. 4.4b, the following sounds can be distinguished (Walker et al. 1990; Kaniusas 2007):

- The *third heart sound*: it is associated with the *rapid ventricular filling* (i.e., *passive* ventricular filling) when the *compliant* ventricular walls (mainly the left ventricular walls) twitch and generate sounds. This sound occurs in the early diastole after the second heart sound (Fig. 4.4b); compare with the rapid filling phase from Fig. 2.38b. It is relatively short and includes spectral components of *very low frequency* in the range of 25–50 Hz because walls are relaxed. The third heart sound is also known as *ventricular gallop* which refers to the cadence of the three heart sounds (i.e., first, second, and third sounds), occurring in rapid succession.
- The *fourth heart sound*: it is associated with the *contraction of atria*, displacing blood into ventricles (i.e., *active* ventricular filling). If the ventricular *compliance is decreased* (abnormal on the left or right side of the heart), the blood strikes the ventricles which start to vibrate and produce audible sounds. This sound occurs in the late diastole just before the first heart sound (Fig. 4.4b). It includes spectral components of *very low frequency* in the range of 20–30 Hz. The fourth heart sound yields an auscultatory cadence that resembles the canter of a horse, thus this sound is also designated as *atrial gallop*.
- *Ejection sounds* (or ejection clicks): they are associated with the (maximal) *opening of semilunar valves*, in which a sudden tensing or an abrupt opening of valves generates sounds. Also the *rapid distention of the aortic root or pulmonary artery* (abnormally dilated) at the onset of ejection may contribute to these sounds (Fig. 4.3). In particular, these sounds arise when either the aortic or pulmonary valve is diseased (e.g., stenotic valves are present). The ejection sounds occur shortly after the first heart sound with the onset of ventricular ejection during *systole* (Fig. 4.4b). The sounds are *high frequency* clicky sounds.
- *Opening sounds* (or opening snaps): constitute the diastolic correlate of the ejection sounds. They are associated with the (maximal) *opening of atrioventricular valves* provided that the mitral or tricuspid valve is diseased and its opening is pathologically arrested. These sounds occur after the second heart sound with the onset of ventricular filling during *diastole* (Fig. 4.4b). The sounds are *high frequency* clicky sounds.
- *Murmurs*: these abnormal sounds are mainly associated with the induced turbulent blood flow in the course of backward regurgitation through leaking valves and forward flow through narrowed or deformed valves (e.g., stenotic valves are present); in addition, vibrations of loose structures within the heart may contribute to murmurs. These sounds may occur either during *systole* (systolic murmurs) or during *diastole* (diastolic murmurs). Murmurs are *high frequency* noisy sounds, whereas the frequency is roughly proportional to the velocity of blood flow; compare the discussion about (4.1) and Footnote 5.

Consequently, heart sounds offer information on *valvular*, *myocardial*, and *hemodynamicactivities*, deterioration of which leaves audible traces within heart sounds.

4.1.1.2 Lung Sounds

Lung sounds comprise another type of body sounds which can be auscultated on the skin (Fig. 4.1). In general, these sounds are related to *turbulentair flow* in relatively large airways, located outside and inside the lungs (Fig. 4.6). *Vibrations of the air andairway walls* are induced, which propagate through the lung tissue (lung parenchyma) and the thoracic tissue towards the skin, the site of the sound's auscultation. Lung sounds yield direct information on the dynamics and ventilation of the *upper airways*, e.g., for the diagnosis of apneas (Sect. 3.1.2), and on the dynamics and ventilation of the *lower airways*, e.g., for the diagnosis of asthma (an inflammatory disease of airways).

In contrast to heart sounds (Sect. 4.1.1.1), *lung sounds* are much more *versatile* and *variable over time*. The status of the lung sounds nomenclature can be best viewed by the Laennec's notice—Laennec is the inventor of the stethoscope (Sect. 1.2.1)—that lung sounds heard were easier to distinguish than to describe (Sect. 1.2.2). Ironically, some physicians devaluated lung sounds as "the sound repertoire of a wet sponge such as the lung is limited" more than 30 years ago (Pasterkamp et al. 1997b).

In particular, lung sounds are generated in the large airways and the lungs (Fig. 4.6). These sounds can be roughly classified into (Dalmay et al. 1995; Loudon and Murphy 1984; Pasterkamp et al. 1997b; Kompis et al. 2001; Hadjileontiadis and Panas 1997a; Fachinger 2003; Kaniusas 2007)

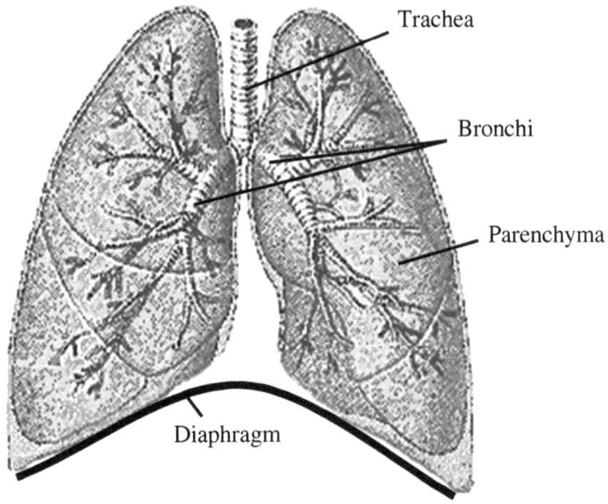

Fig. 4.6 Lungs and adjacent airways relevant for the generation of lung sounds

- normal lung sounds and
- abnormal (adventitious) lung sounds.

The common *classification* of the normal lung sounds is based on the *location of the sound's source*. In particular, the following *normal lung sounds* can be distinguished:

- *Tracheobronchial sounds*: these are the sounds originating in bronchial and tracheal tracts and are heard over or close to large airways (with > 4 mm in diameter), e.g., heard at the chest over the trachea, over the larynx, or on the lateral neck. The sound's source is *centrally situated* and is given by the *turbulent air flowin the bronchi and trachea*, i.e., in the upper airways. The turbulences are due to high *velocity of the air flow*,[5] causing *vibrations of the air and airway walls* during *inspiration* and *expiration*. The propagation distance of sounds towards the skin is relatively short so that sounds, particularly those originating in the trachea, experience only a relatively *weak damping*. As a result, the emitted sounds (at the skin level) are relatively loud and have a tubular (hollow) sound quality as if the air was blown through a tube. The sounds contain frequency components of up to 1 kHz while the component's amplitudes reach baseline levels in the range of 1.2–1.8 kHz (Dalmay et al. 1995); likewise, frequency components are in the approximate range of 100–1,000 Hz (Fachinger 2003). The tracheobronchial sounds are abnormal when heard further away from large airways; it would indicate a consolidation

[5] Streamlined flow or smooth *laminar flow* occurs when air tends to move in parallel layers as if adjacent layers would slide past one another *without lateral mixing*. In a *tube*, the air travelling at the same velocity will be symmetrically arranged around the tube axis, forming cylindrical lamina; the maximum velocity arises at the centre of the tube (Sect. 2.5.2.2). However, the laminar flow can be maintained when it is sufficiently slow or it happens on a sufficiently small scale. Otherwise, rough *turbulent flow* occurs with eddies *leading to lateral mixing* and not contributing to the volume flow rate. The onset of the turbulent flow is roughly determined through the tube's geometry and the *Reynolds number*

$$R = \frac{<u> \cdot 2r \cdot \rho}{\mu}.$$

Here $<u>$ is the average flow velocity of air—with its density ρ and its dynamic viscosity μ—through the tube with the radius r. In an approximation, the turbulent flow starts to develop with $R > 2000$.

Interestingly, the *pressure gradient* scales linearly with the volume *flow* rate in the case of the *laminar flow*, as shown in (2.18). In contrast, the *pressure gradient* is approximately proportional to the *square of the flow* in the case of the *turbulent flow*. Because of the lateral mixing and vortices formed in the turbulent flow, extra energy is required (i.e., disproportionately higher pressure gradient) to maintain the increased movement of air that does not directly contribute to the net flow.

Likewise, the air flow of low u is *laminar* and thus is *silent*. With increasing u (or r) *turbulences* start to occur causing *vibrations* of air and airway walls, which constitute *sound sources*. The arising sound is a *noise-like* signal with a relatively *wide spectrum*, whereas the particular frequency range of noise depends on the level of u.

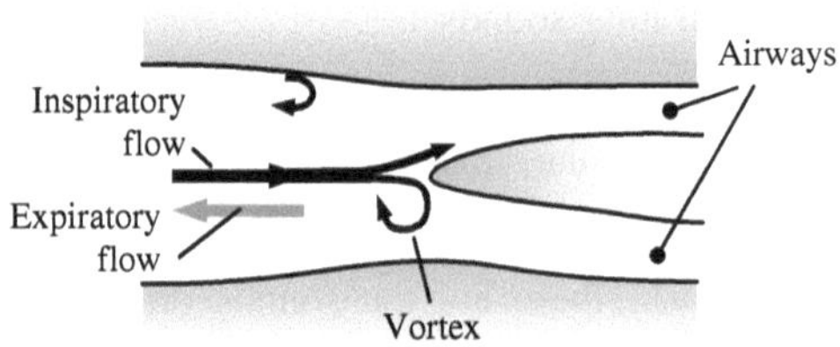

Fig. 4.7 Generation mechanisms of vesicular sounds, i.e., normal lung sounds. Narrowing and branching of small bronchial airways in the lungs constrict and hinder inspiratory air flow, inducing local air turbulences. These turbulences comprise local and distributed (diffuse) sources of lung sounds

of the lung tissue (lung disease) because the consolidation facilitates the sound propagation (i.e., reduces damping of sounds, Sect. 4.1.2.2).

- *Vesicular sounds*: these sounds are heard at sites which are distant from large airways; e.g., at the chest in the peripheral lung fields. The sound's sources are *distributed* throughout the lungs and originate in *air turbulences along bronchi* outside alveoli.[6] The air turbulences are induced by *branching* and *narrowing* of airways, leading to directional changes of the local air flow. The inner surface of airways is *uneven*, which also contributes to the turbulences. In particular, vesicular sounds mainly originate when the air moves into increasingly smaller airways (towards alveoli) during *inspiration*. As shown in Fig. 4.7, airways branch into smaller and smaller airways, whereas the inspiratory air flow hits these branches and air turbulences are created. In contrast, the air moves into increasingly larger airways during expiration, in the course of which the air flow is less confined and it has only a loose contact with the surface of airways. Consequently, less turbulences are created and thus *less sound is generated during expiration* (Fig. 4.7). Vesicular sounds propagate through alveolar tissue (lung parenchyma) towards the skin, experiencing a relatively *large damping*. These emitted sounds (at the skin level) have a soft sound quality as if the air was blown through leaves of a tree. The sounds contain frequency components clearly distinguishable at about 100 Hz (in the range of 100–400 Hz (Fachinger 2003)); the amplitude's *fall-off to baseline levels* at about 1 kHz is much more rapid than for tracheobronchial sounds (Dalmay et al. 1995). In contrast to tracheobronchial sounds, vesicular sounds show *lower intensity* and *narrower spectral range*; e.g., the frequency components above 1 kHz were more clearly auscultated over the trachea than at the chest (Loudon and Murphy 1984). In fact, these differences are due to a strong filtering of vesicular sounds when

[6] In *alveoli* the velocity of the air flow is very low because of a very large total cross-sectional area of the airways. Consequently, the *air flow is laminar* and air turbulences are missing; compare Footnote 5.

propagating to the chest skin. Vesicular sounds tend to have longer transmission paths with more damping (inertial) components involved.

- *Bronchovesicular sounds*: these sounds are intermediate in their characteristics between tracheobronchial and vesicular sounds.

Figure 4.8 illustrates *normal lung sounds* auscultated on the *neck* in comparison with the sounds from the *chest*. Tracheobronchial sounds dominate on the neck (Fig. 4.8a) while vesicular sounds dominate at the chest (Fig. 4.8b). It can be observed that lung sounds can not be recognised in the time domain, only *heart sounds* (Fig. 4.5a) are easily discernable here because of their relatively high intensity. For instance, heart sounds are approximately 30 dB stronger than lung sounds if auscultated on the chest, as discussed in Sect. 4.2.2.3. However, the discussed behaviour of tracheobronchial and vesicular sounds is disclosed in the corresponding *spectrograms*. In fact, the *non-linear logarithmic scaling* (dot color) in the spectrograms accentuates *weak lung sounds* in the presence of *strong heart sounds*. As expected, *tracheobronchial* sounds (Fig. 4.8a) occur during both *inspiration and expiration*, whereas *vesicular* sounds (Fig. 4.8b) dominate only during *inspiration*. It can also be observed that tracheobronchial sounds include a wider range of *frequency components up to about 1 kHz* in comparison to vesicular sounds with components *up to about 500 Hz*, which is in good agreement with the sound descriptions from above.

The *abnormal lung sounds* (adventitious lung sounds) are heard in pathological cases only. Their common classification is based on the *sound's duration*, i.e., continuous sounds with a duration of more than 250 ms can be distinguished from discontinuous sounds with a duration of less than 20 ms (Loudon and Murphy 1984; Pasterkamp et al. 1997b; Rappaport and Sprague 1941; Hadjileontiadis and Panas 1996, 1997b; Iyer et al. 1989; Mikami et al. 1987).

- Abnormal *continuous sounds*: these sounds extend over a relatively long period of time of more than 250 ms and have musical character. A further subdivision is commonly used:
  - *Wheezes*: they seem to arise in the course of interaction between walls of central and lower airways and, on the other hand, the air flow passing these airways. In particular, *narrowed and constricted airways* favour *elastic oscillation* of airway walls (i.e., gradual opening and closure of airways in the radial direction) provided that the *air flow* is limited and non-zero at the constricted site; compare Sect. 4.1.1.3 and Fig. 4.12a. In extreme cases, the narrowing can go to the point where opposite walls touch each other. Arising *vibrations* of the air and airway walls as well as induced *turbulences* contribute to wheezes.[7] These sounds are *high frequency* sounds resembling musical noise.

[7] A very short musical wheeze is also known as *squawk*. Squawks are a combination of wheezes and crackles; they are thought to occur from an explosive opening of airways and fluttering of unstable airway walls.

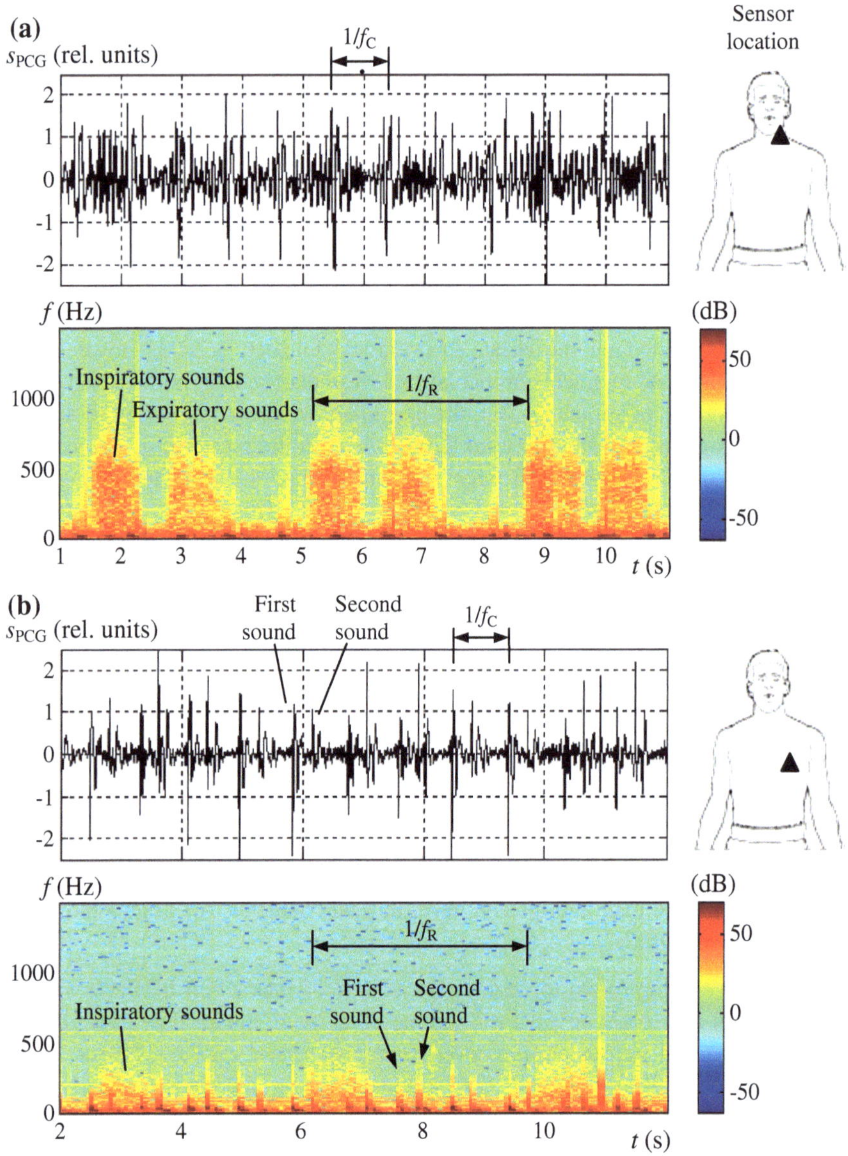

Fig. 4.8 Normal lung sounds while breathing at rest. (**a**) Tracheobronchial sounds, as illustrated by an acoustic biosignal phonocardiogram s_{PCG} from the neck region. (**b**) Vesicular sounds, as illustrated by s_{PCG} from the chest region. The heart rate f_C and respiratory rate f_R are indicated. The corresponding spectrograms (*lower subfigures*) show differences between the tracheobronchial and vesicular sounds. For parameters of the spectrograms see Footnote 4

- *Rhonchi*: these sounds originate in the relatively large airways such as bronchi or bronchioles which are *partially obstructed.* The narrowing is usually due to excessive mucous *secretions* or local swellings. An increased velocity of air flow through thick mucous secretions and the rupture of fluid films contribute to the sound's generation, similar to the generation mechanism of wheezes. Rhonchi are *low frequency* sounds and have a sonorous snoring sound quality.
- *Stridors*: these are intense monophonic wheezes indicating *obstruction of the upper airways* such as of trachea or larynx.

- Abnormal *discontinuous* sounds: these sounds extend over a relatively short period of time of less than 20 ms and have implosive noise-like character. These are due to *explosive reopening* of small airways or fluid-filled alveoli during respiration, previously closed by excessive fluid or lack of aeration. In addition, *bubbling* of the air through copious secretions may contribute to discontinuous sounds. In all cases a rapid equalisation of gas pressures (downstream and upstream) and a rapid release of tensions of airway walls occur; it results in a series of distinct vibrations of the air and airway walls, generating intermittent explosive sounds. Excessive accumulations of secretions in airways or diseased lungs (e.g., inflammation or swelling in tissues surrounding airways) yield discontinuous sounds which can be further subdivided into:
  - *Coarse crackles*: these sounds exhibit a relatively long duration and *low frequency* components, and are mainly indicative of large fluid accumulation and bubbling of the air.
  - *Fine crackles*: in contrast to coarse crackles, fine crackles show short duration and *high frequency* components, and are mainly indicative of airway reopenings.

Figure 4.9 illustrates *vesicular lung sounds* in more detail, which were auscultated at the *chest* while normally breathing. In the time domain (Fig. 4.9a), vesicular sounds can be hardly recognised, as already mentioned. However, if the time axis is stretched out and the amplitude resolution is increased so that signal details can be examined—as depicted in the upper subfigures of Fig. 4.9a—vesicular sounds become uncovered. During *inspiration*, an oscillation with the fundamental harmonic frequency f_{R1} of about 260 Hz can be observed, whereas at the end of expiration oscillatory contributions are absent; this is in full agreement with the discussed generation mechanisms (Kaniusas et al. 2005).

Likewise, the *cardiac component* s^{C}_{PCG} of the auscultated biosignal (Fig. 4.9a) is almost constant within the zoomed region of 10 ms duration because heart sounds dominate up to about 100 Hz, i.e., 100 Hz = 1/10 ms. Conversely, the *respiratory component* s^{R}_{PCG} of the biosignal can be observed within this time resolution because vesicular sounds go up to about 500 Hz, i.e., numerous oscillatory periods of respiratory sounds fit into 10 ms. The *relative amplitude* of vesicular sounds in relation to that of heart sounds can be easily estimated from Fig. 4.9a, yielding

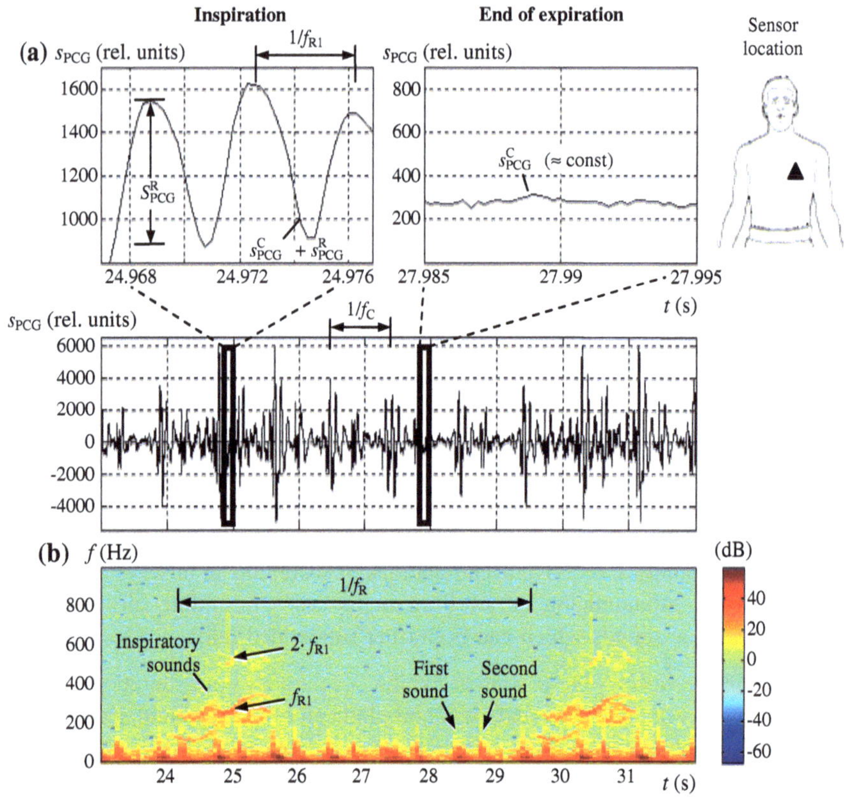

Fig. 4.9 Vesicular lung sounds while breathing at rest. (**a**) Acoustic biosignal phonocardiogram s_{PCG} from the chest region with indicated heart rate f_C. A zoomed region of s_{PCG} with the duration of 10 ms is given at the time of inspiration, including the cardiac component s^C_{PCG} and respiratory component s^R_{PCG} (*left upper subfigure*), in comparison to the end of expiration including only s^C_{PCG} (*right upper subfigure*). (**b**) The corresponding spectrogram with indicated respiratory rate f_R and fundamental harmonic frequency f_{R1}. For parameters of the spectrogram see Footnote 4

about −24 dB (= 20 · log(700/11000)); that is, s^R_{PCG} exhibits much smaller oscillation amplitude than s^C_{PCG}; compare Sect. 4.2.2.3.

Again, vesicular sounds manifest in the *spectrogram* during inspiration (Fig. 4.9b). In particular, the fundamental harmonic with f_{R1} can be seen, the same as already disclosed in the expanded waveforms of Fig. 4.9a. Additionally, a second harmonic appears at $2 \cdot f_{R1}$.

As can be derived from the origin of normal lung sounds, at the *chest wall* expiratory sounds originate from a more central source then inspiratory sounds (Earis 1992; Dalmay et al. 1995). Likewise, *central compact sources* of tracheobronchial sounds (located in the upper large airways) contribute more to the auscultated sounds during *expiration*, whereas *local distributed (diffuse) sources* of vesicular sounds (located in the distal bronchial airways) contribute more during *inspiration*.

In consequence, the inspiratory sounds show a relatively large amplitude (intensity) and include components of high frequency due to close vicinity of the auscultation site to inspiratory (distributed) sound sources. In contrast, the expiratory sounds are relatively weak due to distant (central) sound sources, long sound transmission paths, and thus strong accumulated damping of sounds. For instance, author in Fachinger (2003) report that the inspiratory sounds on the anterior chest showed twice as large intensity as that of the expiratory sounds.

An important characteristic of *normal lung sounds* is that their spectra are clearly linked to the strength of the respiratory *air flow* q^{A} (Dalmay et al. 1995). In an approximation, the sound intensity or the *lung sounds amplitude* S_{PCG}^{R} of normal lung sounds—as illustrated in Fig. 4.9a—increases exponentially with increasing q^{A}, that is

$$S_{PCG}^{R} = c \cdot \left(q^{A}\right)^{\alpha}. \tag{4.1}$$

Here c is a positive constant and α is another constant representing the power index of q^{A}. In the case of

- *tracheobronchial sounds*, frequency components shift upward in frequency and their amplitudes increase in proportion to q^{A}, this relationship being more marked during *inspiration* (Dalmay et al. 1995). Likewise, $\alpha = 1$ applies in (4.1).
- The regional intensity of *vesicular sounds* correlates with increasing level of regional ventilation (Loudon and Murphy 1984; Jones et al. 1999). Similarly, the amplitude of vesicular sounds linearly increases and frequency components shift upwards as q^{A} rises, which is particularly pronounced during *inspiration* (Dalmay et al. 1995). In the case of the *linear relationship*, $\alpha = 1$ applies; however, *non-linear relationships* were also reported with $\alpha = 1.75$ (Fachinger 2003) or $\alpha = 2$ (Pasterkamp et al. 1997b; Earis 1992).

It should be noted that reduced *intensity of vesicular sounds* was reported to be a strong indicator of *obstructive* pulmonary disease (Pasterkamp et al. 1997b), i.e., to be an indicator of impaired local ventilation of the lungs. In contrast, an increase in the intensity of vesicular sounds is considered indicative of *lung expansion* (Jones et al. 1999). For the *typical intensity levels* of normal lung sounds see Sect. 4.1.1.3.

Lastly, *large variability of lung sounds* should be addressed in some depth (Dalmay et al. 1995; Jones et al. 1999; Kompis et al. 2001). The content of lung sounds in the time and frequency domain—such as demonstrated in Fig. 4.9—greatly depends on the particular site of auscultation within *one subject* (Sect. 4.2.2.3), the degree of voluntary control that the subject is able to exert over breathing, body position (e.g., sitting or lying), and, of course, the actual respiration phase, i.e., inspiration or expiration. The temporal variability of lung sounds is more pronounced during expiration than inspiration (Dalmay et al. 1995). In fact, this variability is mainly due to strong influence of individual morphology of airways (Pasterkamp et al. 1997b) and lung-muscle-fat ratios (Kompis et al. 2001). Lung sounds vary greatly *among subjects* ventilating even at similar q^{A} (4.1),

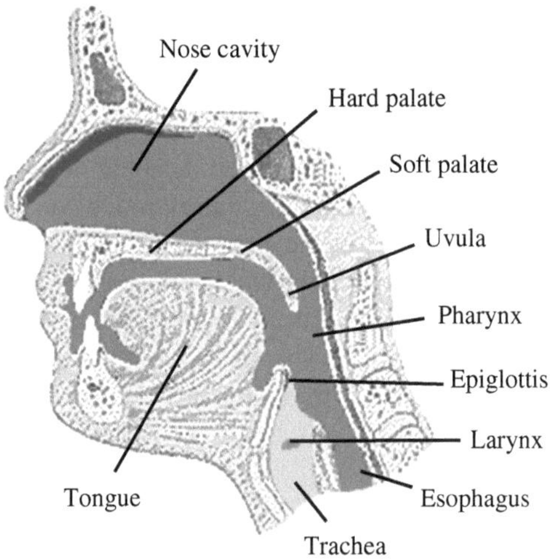

Fig. 4.10 Pharyngeal airways and surrounding structures of the upper airways relevant for the generation of snoring sounds

whereas the variability is still considerably high after introduction of corrections for diverse physical characteristics of subjects such as body weight, body height, the subject's age, and body surface area.

4.1.1.3 Snoring Sounds

While heart and lung sounds have been in the focus of clinical investigations for centuries, only recently medical interest has focussed on *snoring sounds*[8] while sleeping (Fig. 4.1). In general, these sounds are related to *vibrations of instable structures* in the upper airways (such as the soft palate or uvula), *radial oscillation* of (pharyngeal) *flexible airway walls* and oscillatory *narrowing* of airways (up to their complete occlusion), and *turbulences of the air*[9] (Fig. 4.10). Snoring sounds

---

[8] Epidemiological studies have shown that nearly 40 % of males and about 20 % of females are snorers (Saletu 2001). The *prevalence* of habitual snoring rises markedly after the age of 40, whereas more than 60 % of males and more than 40 % of females are snorers in this aged population (Beck et al. 1995).

[9] In fact, the physics of sound formation in *snoring* is very similar to that in *speech* (Perez-Padilla et al. 1993). For instance,

- voiced sounds are related to *vibrations* of vocal cords,
- fricative sounds are related to the friction of *turbulent air flow* through a narrow orifice, and
- explosive sounds are related to sudden *release of pressure*.

are favoured by various *physiologic and social factors*,[10] whereas these sounds yield direct information on the dynamics and ventilation of the *upper airways*, e.g., for the diagnosis of apneas (Sects. 3.1.2 and 4.1.1.4). Snoring could be related to sleep deprivation and thus to other *severe pathologies*.[11]

In particular, the *snoring is proceeded* by a temporal decrease in the *diameter* of the oropharynx (Fig. 4.10), which can be even reduced to a slit[12] (Liistro et al. 1991; Cirignota 2004). Figure 4.11a, b demonstrates the narrowing of the pharynx by video images. Likewise, the (supraglottic) *resistance of the airway* to the air flow increases—by a factor of about 3, estimated from Liistro et al. (1991)—which in the case of heavy and obstructive snoring (i.e., *spontaneous snoring* during sleep) leads to initial *flow limitation* before onset of the snoring. In particular, the flow first increases as the driving pressure increases but then it saturates, i.e., the air flow becomes limited.

As the *snoring begins* and continues—mainly during *inspiration* (Liistro et al. 1991)—fluttering of loose structures in the upper airways occur, especially *vibrations* of the soft palate and pharyngeal walls (Perez-Padilla et al. 1993; Beck et al. 1995). Likewise, the appearance of *repetitive and steady sound* structures in the time domain during snoring coincides with the time course of airway wall motions and the time course of the air flow oscillation.

---

[10] *Snoring* is favoured by *physiological factors* such as small pharyngeal area and increased pharyngeal floppiness, i.e., excessive change in pharyngeal area occurs in response to applied air pressure (Saletu and Saletu-Zyhlarz 2001; Brunt et al. 1997). In addition, the supine sleep posture (i.e., retroposition of the tongue), obesity (i.e., high body mass index *BMI*, Footnote 202 in Sect. 3), large neck circumference (Sergi et al. 1999), presence of space occupying masses which block airways (e.g., hypertrophy of the soft palate or uvula), or a pathological narrowing of the nasal airway facilitate (disadvantageously) the generation of snoring. Among *social factors* contributing to the occurrence of snoring, mental stress, tiredness, and alcohol intake can be mentioned. Interestingly, *subjective factors* as familiar home settings or less familiar sleep labs also seem to influence the severity of nocturnal snoring which, in fact, tends to be heavier while sleeping in a sleep lab (Series et al. 1993).

[11] From the *physiological* point of view, the snoring, especially *obstructive snoring*, may be connected to increased morbidity, systemic hypertension, cerebrovascular disease, stroke, and even impaired cognitive functions (Saletu and Saletu-Zyhlarz 2001; Series et al. 1993; Wilson et al. 1999). In addition, obstructive and loud snoring is a major cause of disruption to other family members besides the snorer himself; it represents a disadvantageous *social* impact of snoring.

[12] The *narrowing of the pharyngeal airway* (or even its partial and *passive collapse*) can be due to negative oropharyngeal pressure generated during *inspiration*, relaxation of the pharyngeal muscles, or even sleep-related fall in the tone of the upper airway muscles (Liistro et al. 1991). The pharyngeal muscle tone is reduced not only during sleep, but also under the influence of alcohol or drugs (Saletu and Saletu-Zyhlarz 2001).

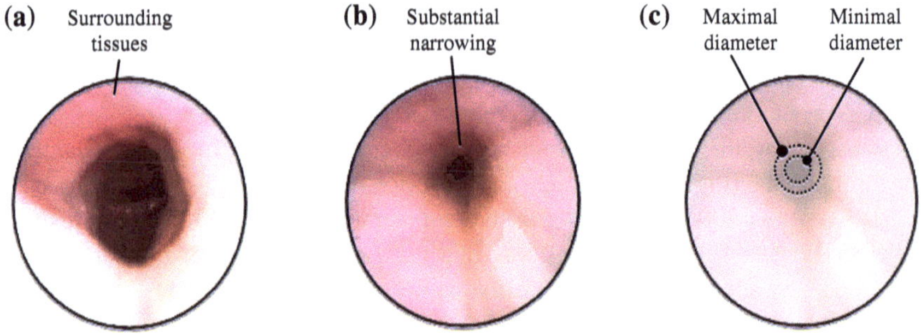

Fig. 4.11 Video images of the pharynx as recorded from the mouth cavity (Hohenhorst 2000). (**a**) Before inspiration. (**b**) During inspiration. (**c**) During inspiration with schematically indicated oscillation of the airway diameter

Figures 4.11c and 4.12a illustrate such *elastic oscillations* of airway walls provided that a local narrowing (*flow-limiting segment*) is present within the depicted *highly compliant airway*. *Aeroelastic interactions* occur between the air flow and the airway wall. At the constriction, the air flow is confined to a smaller cross sectional area A (Fig. 4.12a). Consequently, the velocity u of the air flow increases at the constriction site ($u_2 > u_1$) because the net flow

$$q^{\mathrm{A}} = <u> \cdot A \tag{4.2}$$

does not change along the airway (i.e., q^{A} is assumed to be constant); compare (2.17). The lateral pressure p of the air flow must correspondingly decrease ($p_2 < p_1$),[13] *favouring the narrowing of the constriction* even more. Likewise, radial forces keeping the airway open are reduced and thus there is an *inward* swing of airway walls[14]; the constricted site becomes more pronounced and unstable (collapsible) at

[13] The *Bernoulli's equation* governs the behaviour of u and p in an ideal flow, which is deduced from the principle of the conservation of energy (Nichols and O'Rourke 2005). According to Fig. 4.12, the total energy, i.e., the sum of potential and kinetic energies, at a non-constricted site with p_1 and u_1 is equal to the total energy at a constricted site with p_2 and u_2, considering a single horizontal airway. It yields

$$p_1 - p_2 = \frac{1}{2} \cdot \rho \cdot \left(u_2^2 - u_1^2 \right) ,$$

where ρ is the air density. The latter equation demonstrates that $p_2 < p_1$ if $u_2 > u_1$, i.e., p is decreased at the constricted site if a constant q^{A} along the airway is given. In other words, the opening pressure at the constricted site is decreased, which promotes the airway collapsibility at its constriction even more.

[14] It seems that the critically *low cross-sectional area* and critical *limitation of the flow* q^{A} initiate the oscillation of airway walls (Liistro et al. 1991; Perez-Padilla et al. 1993). In general, the limitation of q^{A} appears when u equals the velocity of propagating pressure pulse waves along the airway. Likewise, the oscillations occur more readily at a lower q^{A}, provided that the *compliance of the airway* is high.

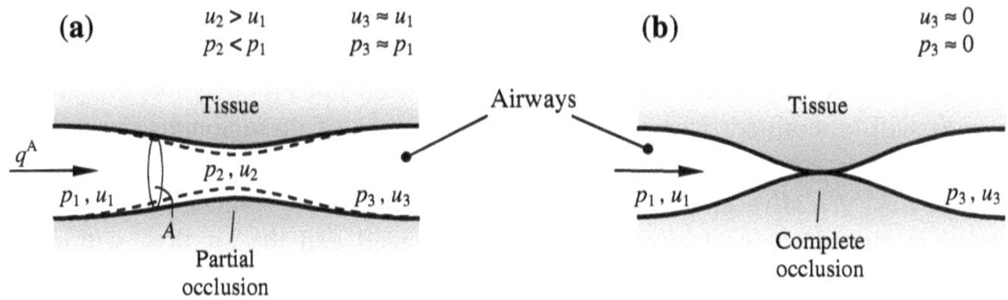

Fig. 4.12 (**a**) Generation of continuous snoring sounds because of an oscillatory narrowing and local constriction of large upper airways; compare widened (*bold*) and narrowed (*dashed*) airways. The relations of the pressure p and velocity u of the air flow q^{A} along the depicted airway are indicated; compare Footnotes 13 and 15. (**b**) Generation of obstructive snoring sounds due to a repetitive and temporal occlusion of large upper airways. The walls of the airways show a high compliance with excessive masses involved, which favours the collapsibility of airways and reduces their permeability to air; compare Footnote 16

$q^{A} \neq 0$. However, *elastic forces* appear progressively in airway walls, which are directed *outwards* and cause the deflected walls to swing back to their neutral position; an oscillatory vibration of airway walls results[15]; the latter mechanism dominates also the *generation of wheezes* in obstructed sites (Sect. 4.1.1.2).

On the other hand, a *collapsible airway* experiencing a local pressure decrease (Fig. 4.12a) can even collapse and completely occlude the lumen for a brief period of time (Fig. 4.12b). In particular, large amplitude oscillations of airway walls can yield partial or complete occlusion of the airway, with the point of maximum constriction moving upstream along the airway. Repetitive openings of the occluded airway generate *abrupt pressure equalizations* (popping openings) and *tissue vibrations*, which emit series of *explosive and discontinuous sound* structures—reoccurring with the frequency of the openings—in the time domain.[16]

Further narrowing of the oropharynx during snoring may lead to even louder snoring and laboured breathing. In extreme cases, progressive narrowing can yield a *sustained and complete occlusion* of the upper airway, which then manifests as the obstructive *sleep apnea* (Sect. 3.1.2).

---

[15] This theory is called "*flutter theory*" (Perez-Padilla et al. 1993); compare Footnote 16. That is, it explains the *continuous form* of snoring sounds in the time domain. These sounds arise in the course of oscillations of airway walls when the airflow is forced through a highly compliant airway and can interact with the elastic walls (Fig. 4.12a). The resulting *oscillation frequency* tends to decrease with increasing wall thickness and decreasing longitudinal tension in the walls, this tension being also affected by the activity of pharyngeal muscles.

[16] This theory is called "*relaxation theory*" (Perez-Padilla et al. 1993); compare Footnote 15. That is, it explains the *discontinuous form* of *explosive snoring sounds* in the time domain, which are due to repetitive openings of local occlusions of the airway (Fig. 4.12b). The resulting *oscillation frequency* is relatively low because of *large radial deflection* of airway walls.

In addition, local *turbulences of the air* in the upper airways seem to contribute to the emission of snoring sounds; compare Sect. 4.1.1.2. That is, the (increased) level of u in the (narrowed) airway (Fig. 4.12a) determines if *noise-like broadband sounds* will be emitted, whereas the frequency range of these sounds depends on this level of u; compare Footnote 5.

Likewise, *fricative turbulent quality* of snoring sounds is related to *air turbulences* in the pharyngeal airway which is narrowed and the air flow within the airway is limited at a lower value than during rattling snoring (Perez-Padilla et al. 1993). In contrast, *regular rattling quality* of snoring sounds is due to *oscillating structures* such as the pharyngeal walls or soft palate. Relatively small oscillations yield steady continuous waveforms (according to the "flutter theory", Footnote 15) while relatively large oscillations with recurring reopenings yield series of repetitive explosive structures in the sound waveform (according to the "relaxation theory", Footnote 16).

In general, *characteristics of snoring sounds* are mainly determined by the air pressure and air flow in the upper airways in combination with the compliance and collapsibility of airways (Series et al. 1993). Usually the energy of snoring sounds is limited to their frequency components *below 2 kHz* (Perez-Padilla et al. 1993).

Like lung sounds (Sect. 4.1.1.2), *snoring sounds* show high diversity and are subjected to large *variability*. The footprints of snoring sounds in the time and frequency domains can even change from one breath to the other. Consequently, there are numerous possibilities to classify snoring sounds. The most commonly used *classifications* are based on (Kaniusas 2007)

- the location of the sound source,
- the diagnostically relevant type of snoring, and
- the waveform of snoring sounds in the time domain.

In accordance with the location of the *sound source*, the following snoring sounds can be distinguished (Liistro et al. 1991):

- *Nasal snoring*: during simulated nasal snoring (breathing exclusively through the nose) the resulting mainly inspiratory sounds originate in the course of the *uvula* vibrations (Fig. 4.10) while the soft palate and the back of the tongue remain in close contact (Liistro et al. 1991). On the other hand, spontaneous nasal snoring is also due to the *palate* or *pharyngeal wall* vibrations (Perez-Padilla et al. 1993). For instance, the oscillation frequency of the uvula is about 80 Hz (Liistro et al. 1991). In the frequency domain, the nasal snoring shows discrete sharp peaks below 500 Hz, i.e., a peak at a fundamental harmonic frequency and subsequent peaks at its harmonics. These peaks correspond to the resonant peaks (formants) of the *airway's resonating cavities* and suggest that a *single sound source* dominates nasal snoring[17] (Perez-Padilla et al. 1993).

---

[17] In general, *spectral characteristics* of emitted sounds result from both the *source of sound* and *filtering properties* (or resonant properties) of the airway (Perez-Padilla et al. 1993). For instance, the *source properties* change when a different segment starts to oscillate or the mechanical

- *Oral snoring*: this snoring through the open mouth (breathing exclusively through the mouth) is characterized by vibrations of the whole *soft palate* and a dramatic decrease in the cross-sectional area of the oropharynx (Fig. 4.10), yielding a lower oscillation frequency of about 30 Hz in comparison with the nasal snoring (Liistro et al. 1991). This difference may be attributed to a larger oscillating mass of the soft palate than that of the uvula, yielding a lower oscillation frequency for a larger oscillating mass.
- *Oronasal snoring*: these snoring sounds (breathing through the nose and mouth) exhibit a mixture of sounds similar to *nasal snoring* and fricative *noisy sounds* characteristic of a source in the *turbulent flow* (Perez-Padilla et al. 1993); compare Footnote 5. Likewise, in the frequency domain a mixture of discrete sharp peaks and broad-band (white) noise dominate in the range of up to about 1,300 Hz. The large number of peaks may reflect two or more segments (e.g., the *uvula* and *soft palate*) oscillating with different frequencies. Figure 3.6c demonstrates an oscillation of the air flow during snoring with the rate of about 40 Hz.

Considering the diagnostically relevant *types of snoring*, the following snoring sounds can be distinguished (Liistro et al. 1991; Perez-Padilla et al. 1993; Series et al. 1993; Beck et al. 1995):

- *Normal snoring*: spontaneous snoring is always preceded by the *limitation of the air flow* and the *narrowingof the pharyngeal airway* (Liistro et al. 1991; Perez-Padilla et al. 1993; Series et al. 1993). The oscillation of airway walls occurs in line with the "*flutter theory*" (Footnote 15), as illustrated in Fig. 4.12a. The supraglottic pressure[18] if depicted over q^{A} forms a *hysteresis loop*, i.e., oscillations of the supraglottic pressure and q^{A} are 180° out-of-phase. The latter behaviour can be explained by consecutive (partial) closings and openings of the pharynx by the soft palate, yielding opposite changes in the supraglottic pressure and q^{A} (Liistro et al. 1991). Normal snoring sounds show a *regular rattling character* (Perez-Padilla et al. 1993) with dominant frequency components in the range of 100–600 Hz and minor components up to 1 kHz (Beck et al. 1995). Spectral peaks in the frequency domain occur at regular distances and represent *harmonic* waves.

(Footnote 17 continued)
characteristics of the other *oscillating segment* are different. In analogy, the *filtering properties* change when geometric dimensions of the pharynx or mouth cavity vary over time (i.e., dimensions of resonating cavities in front of the source location, cavities acting as *band-pass filters*; compare Fig. 4.24), or dimensions of neighbouring apertures for the air escape vary over time. In fact, *emitted sounds* are determined by a *product of the sound source* (usually broadband source) and the *filtering function* (band-pass filters) of the airway.

[18] *Supraglottic pressure* is the pressure drop along the upper airway above the epiglottis, see Fig. 4.10.

- *Obstructive snoring*: this pathological type of snoring (dominating in humans with the *obstructive sleep apnea*, Sect. 3.1.2) is associated with *repetitive temporal occlusion* and opening of a strongly *narrowed and collapsible airway*. This is due to high compliance of airway walls, as illustrated in Fig. 4.12b, and corresponds to the "*relaxation theory*" (Footnote 16). In addition, obstructive snoring is related to oscillations of the *soft palate*. The hysteresis loop is larger in size compared to normal snoring. Obstructive snoring sounds are *louder* than normal snoring sounds and have *fricative high-pitched quality*. In the time domain, the sounds exhibit *intermittent bursts of noise* at regular intervals—related to the fundamental harmonic frequency—and dramatically variable sound patterns. Typically, regular discrete peaks, i.e., spectral *harmonics*, can be observed in the frequency domain with the frequency components extending up to 2 kHz; compare Sect. 4.1.1.4. Obstructive snoring sounds have a higher cumulative power above 800 Hz relative to the power below 800 Hz when compared to normal snoring (Perez-Padilla et al. 1993). In an approximation, the intensity of frequency components decreases with increasing frequency less strongly in obstructive snoring than in normal snoring.
- *Simulated snoring*: this kind of an intentionally provoked snoring is *not preceded by flow limitation*, even though preceded by an increase in the supraglottic resistance of the airway, at variance with the spontaneous snoring during sleep (Liistro et al. 1991). The narrowing of the pharyngeal airway is probably produced by voluntary contraction of the pharyngeal constrictor muscles. The presence of the *hysteresis loop* has also been reported (Liistro et al. 1991). Simulated snoring sounds resemble complex-waveform snoring (as described below) with multiple equally-spaced peaks of power in the frequency domain ranging up to 800 Hz (Beck et al. 1995).

If the distinct *waveform patterns* of snoring sounds in the time domain are taken as the classification basis, the following snoring sounds can be distinguished (Beck et al. 1995):

- *Simple-waveform snoring*: this type of snoring shows a nearly *sinusoidal waveform* or a *periodic waveform* (with a secondary deflection) over time with minor secondary oscillations. Consequently, only one up to three equally-spaced peaks (i.e., only a few harmonics) dominate in the frequency domain in the range of about 100–240 Hz while the first peak (at the lowest frequency) is usually the most prominent. The simple-waveform snoring probably results from the *vibration of airway walls* around their neutral position without the actual closure of the lumen (compare Fig. 4.12a).
- *Complex-waveform snoring*: it is characterized by *repetitive, equally-spaced segments* in the time domain, whereas each segment starts with a large deflection and ends with a decaying wave. The repetitive segments arise with the frequencies in the range of about 60–140 Hz, thus the frequencies are lower than in the simple-waveform snoring. More rapid, secondary *oscillations within each*

segment occur in the range of up to about 1,000 Hz. Therefore, a comb-like structure with multiple peaks of different amplitudes can be observed in the frequency domain, whereas the frequency interval between the peaks is equal to the arousal frequency of the repetitive segments. In contrast to the simple-waveform snoring, the complex-waveform snoring probably results from *repetitive collisions of airway walls* with intermittent brief closures of the lumen.[19] It should be noted that the simple-waveform and complex-waveform snoring can be found in a single subject, whereas one type of snoring can change to the other even within a single snore.

Figure 4.13 illustrates *normal snoring sounds* in more detail, which were auscultated at the *chest* during sleep. Provided that the snoring is relatively *silent* (Fig. 4.13a), snoring sounds can be hardly recognised in the time domain (Kaniusas et al. 2005); only heart sounds are easily discernable here (compare Fig. 4.9). However, if the time axis is stretched out during an *inspiratory snoring* event and the amplitude resolution is increased—as depicted in the right upper subfigure of Fig. 4.13a—the waveform pattern of snoring sounds becomes uncovered. That is, a *periodic waveform* with the fundamental harmonic frequency f_{R1} of about 160 Hz can be observed. The *relative amplitude* of these snoring sounds in relation to that of heart sounds can be easily estimated and yields –32 dB (= 20 · log(1000/40000)); likewise, the *silent normal snoring sounds* are much *weaker than heart sounds*; compare Sect. 4.2.2.3. Again, normal snoring sounds clearly manifest in the *spectrogram* during inspiration (Fig. 4.13a). In particular, a series of *harmonics* at integer multiples of f_{R1}, i.e., at $k \cdot f_{R1}$ with k as the integer index, can be seen, whereas the fundamental harmonic located at f_{R1} was already disclosed in the right upper subfigure of Fig. 4.13a.

In the case of the relatively *loud normal snoring* (Fig. 4.13b), snoring sounds start to dominate in the time domain (Kaniusas et al. 2005). A nearly *sinusoidal waveform* can be observed in the expanded waveform with $f_{R1} \approx 95$ Hz. The *relative amplitude* of these snoring sounds in relation to that of heart sounds amounts to +2.5 dB (= 20 · log(20000/15000)); likewise, the *loud normal snoring sounds* are already a little bit *stronger than heart sounds*. In the *spectrogram*, the footprint of snoring sounds is more dense and includes more high frequency components as compared with either the relatively silent snoring (Fig. 4.13a) or vesicular lung sounds (Fig. 4.9b).

Specific sound patterns related to *obstructive snoring* are shown in Fig. 4.14, illustrating an evident difference between normal and obstructive snoring. These *snoring sounds* clearly *dominate over heart sounds* in the time domain, whereas the expanded waveform pattern shows *interfering oscillations* with $f_{R1} \approx 85$ Hz (Fig. 4.14a). The *relative amplitude* of obstructive snoring sounds in relation to that

[19] Interestingly, the largest and sharpest deflection of the sound wave coincides with the peak of the air flow, considering the complex-waveform snoring (Beck et al. 1995). It indicates the relevance of the *air flow* for the generation of *snoring sounds*, according to discussed mechanisms shown in Fig. 4.12.

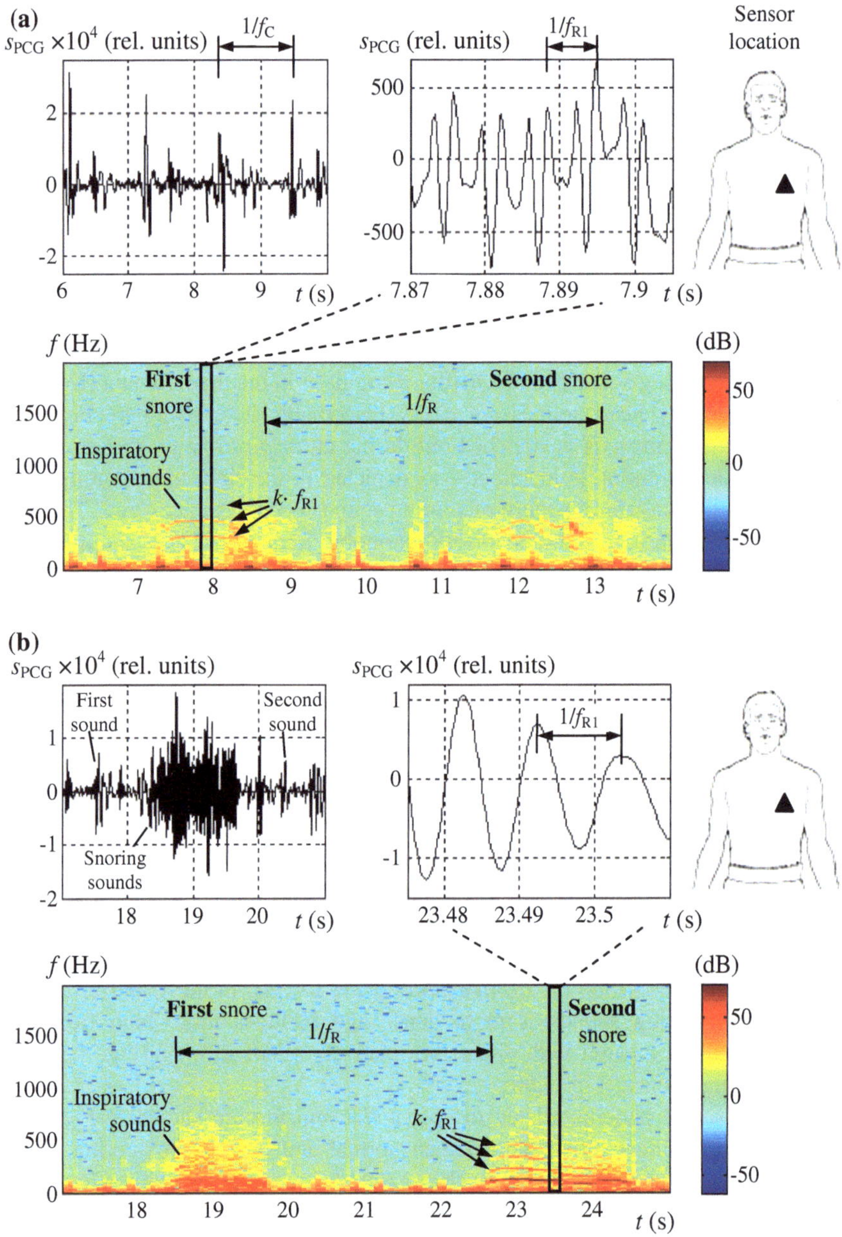
(a)
s_{PCG} ×10^4 (rel. units)
1/f_C
s_{PCG} (rel. units)
1/f_{R1}
Sensor location
t (s)
f (Hz)
(dB)
First snore
Second snore
1/f_R
Inspiratory sounds
k· f_{R1}
(b)
s_{PCG} ×10^4 (rel. units)
First sound
Second sound
Snoring sounds
s_{PCG} ×10^4 (rel. units)
1/f_{R1}
t (s)
f (Hz)
(dB)
First snore
1/f_R
Second snore
Inspiratory sounds
k· f_{R1}

◂**Fig. 4.13** Normal snoring sounds during sleep (compare Fig. 4.14). (**a**) Relatively silent snoring from a male subject (body mass index *BMI* of 23.6 kg/m^2, see Footnote 202 in Sect. 3), as illustrated by an acoustic biosignal phonocardiogram s_{PCG} from the chest region. (**b**) Relatively loud snoring from another male subject (*BMI* = 24.7 kg/m^2). A zoomed region of s_{PCG} with the duration of 35 ms is given in each case at the time of a snoring event (*right upper subfigure*). The heart rate f_C and respiratory rate f_R are indicated. The corresponding spectrograms (*lower subfigures*) with shown fundamental harmonic frequency f_{R1} are given for comparison. For parameters of the spectrograms see Footnote 4

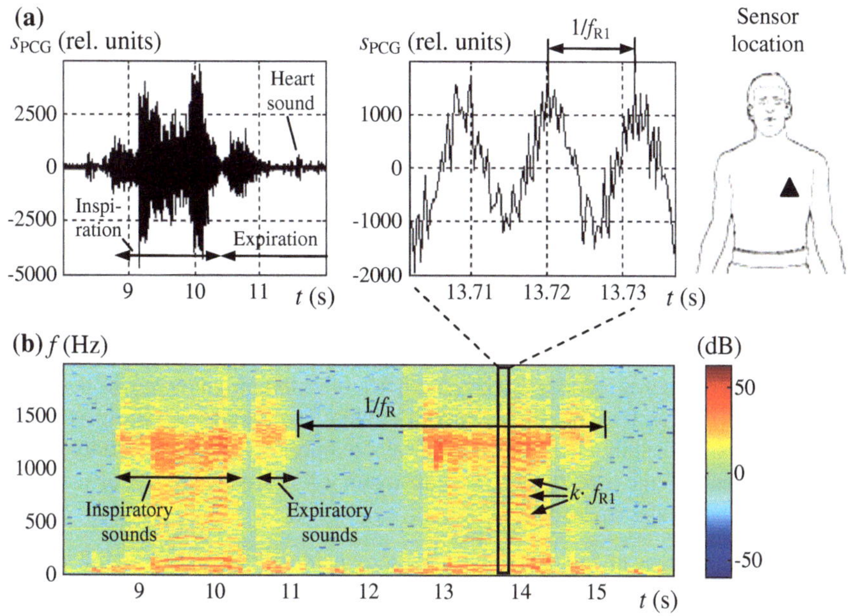

Fig. 4.14 Obstructive snoring sounds during sleep, recorded from a male patient with the obstructive sleep apnea (*BMI* = 35.5 kg/m^2); compare Fig. 4.13. (**a**) Acoustic biosignal phonocardiogram s_{PCG} from the chest region. A zoomed region of s_{PCG} with the duration of 35 ms is given at the time of an obstructive snoring event (*right subfigure*). (**b**) The corresponding spectrogram with indicated respiratory rate f_R and fundamental harmonic frequency f_{R1}. For parameters of the spectrogram see Footnote 4

of heart sounds amounts to +12 dB (= 20 · log(5000/1300)); compare Sect. 4.2.2.3. A series of *harmonics* can be observed in the spectrogram, extending up to 1,500 Hz (Fig. 4.14b). A noise-like structure is visible above 1,000 Hz, which also appears in the course of expiratory snoring sounds but with smaller amplitude. As the dominant feature, the cumulative power of *high frequency components* above 800 Hz seems to be larger than that below 800 Hz.

Simulated snoring sounds are depicted in Figs. 3.5 and 3.6 (Sect. 3.1.2). The corresponding oscillations of the air flow with the frequency $f_{R1} \approx 40$ Hz can be clearly recognised in Fig. 3.6c. In fact, the air flow oscillates with the oscillation frequency of the sound waveform, as also shown in Beck et al. (1995); compare Footnote 19.

Considering the different ways of the snoring sound classification, the illustrated (silent and loud) normal snoring from Fig. 4.13 seems to correspond to the oronasal and simple-waveform snoring. On the other hand, the obstructive snoring from Fig. 4.14 resembles the oronasal and complex-waveform snoring.

In analogy with lung sounds, there are indications that the *amplitude of snoring sounds* depends strongly on—or, as a first approximation, is proportional to—the level of the *air flow* q^A; see Footnote 19. However, this relationship seems to be *non-linear* because different generation mechanisms (i.e., oscillations, reopenings, and turbulences) are involved in the origination of snoring sounds. In addition, q^A is a function of both u and A (4.2) which are important non-linear parameters of these mechanisms.

Similar to lung sounds (Sect. 4.1.1.2), *snoring sounds* are subjected to *strong variability* in their loudness and frequency. The strong variability can be observed in the time and frequency domains, in which sound patterns may change from one snore to another or even experience changes within a single snore (Kaniusas 2007; Moerman et al. 2002; Perez-Padilla et al. 1993). The variability of obstructive snoring is particularly high (Beck et al. 1995). This *variability* in sound patterns may arise due to

- altering (geometric, physical) characteristics of *resonating cavities* in the upper airways such as the pharynx or mouth cavity. The geometry and apertures of cavities significantly change when airways temporarily occlude or fully dilate.
- Movement of the *site of collapse* upstream or downstream the airway also contributes to the sound variability.

The *spectrogram* in Fig. 4.13b clearly demonstrates the *variability* of snoring sounds. In the depicted case, noise-like structure during the first snore (with *fricative quality*) transforms into a series of harmonics during the second snore (with *rattling quality*). Even frequencies of harmonics markedly decrease during the second snore.

Lastly, *intensity levels* of snoring sounds should be shortly discussed, especially in comparison to normal lung sounds (Sect. 4.1.1.2). To begin with, the *background noise* level in rooms could reach 50 dB *sound pressure level* (SPL).[20]

---

[20] The abbreviation *dB SPL* refers to a logarithmic measure of the sound pressure level relative to a reference pressure level (of 20 μPa), i.e., relative to the *threshold of human hearing*. For instance, a normal conversation yields about 60 dB SPL, whereas a pneumatic drill—in a distance of a few meters—yields 100 dB SPL.

However, the human ear does not equally respond to all frequencies and it is highly sensitive to sounds in the frequency range of about 1–5 kHz; likewise, the ear is less sensitive to very low or very high frequencies of sounds. To accommodate this behaviour, sound meters use frequency filters which mimic this non-linear frequency response of the ear. In this context, the abbreviation *dBA* stays for a logarithmic measure of the sound pressure level employing the so-called *A-weighting filter*. This filter disproportionately attenuates very low frequencies, e.g., an attenuation of −30 dB is applied at 50 Hz while no attenuation (of 0 dB) is applied at 1 kHz.

The *lung sound level* is normally in the range of 40–45 dB SPL (Schäfer 1988) or 17–26 dBA (Schäfer 1996), and could go up to 54 dB SPL (Series et al. 1993).

The *snoring sound level* is greater than 60 dB SPL in line with (Series et al. 1993; Itasaka et al. 1999) or greater than 68 dB SPL according to Schäfer (1988), and can temporarily reach values of more than 100 dB SPL in a distance less than 1 m from the head of the snorer (according to diverging reports). The *obstructive snoring* yields levels in the range of 50–70 dBA, the levels increasing by > 5 dBA from nonapneic snoring patients to apneic snoring patients (Wilson et al. 1999). Similarly, *maximum snoring sound levels* of up to 80 dBA were reported in a distance of 1 m for nonapneic snoring patients, whereas maximum levels of up to 94 dBA were reported during postapneic snores in apneic snoring patients (Schäfer 1996). In fact, loud snoring may constitute an excessive noise exposure which can even cause *hearing problems* (Wilson et al. 1999).

4.1.1.4 Apneic Sounds

There is strong evidence that *obstructive snoring* during sleep (Sect. 4.1.1.3) may be an intermediate symptom[21] in the history of the sleep apnea syndrome (Sect. 3.1.2). In particular, the *obstructive sleep apnea* is characterised by a complete occlusion of the upper airways and ceased breathing, as illustrated in Fig. 3.8b. The intermittent respiratory arrest is marked by intermittent absence of breathing sounds, i.e., lung and snoring sounds, which yields a unique *acoustical fingerprint of apneas* among body sounds.

When an obstructive *apnea terminates*, a gasp for the air follows and very loud, high frequency, explosive *apneic sounds* are usually induced by reopening of the collapsed airways at the tongue base. These *inspiratory sounds* appear to be highly different from *regular, rattling snoring sounds in nonapneic subjects* (Perez-Padilla et al. 1993); compare Sect. 4.1.1.3.

Even though snoring sounds greatly vary from one respiration cycle to another, the *first postapneic inspiratory snore* (at the end of an apnea) is distinctive because the *airway is at its narrowest*. That is, instead of rattling and repetitive quality, continuous *turbulent and fricative quality* can be heard such as can be simulated by producing a consonant "h" sound. The *air turbulences* from the air flowing through a *narrow orifice*—with a relatively *high flow velocity* (Footnote 5)—appear to contribute much to the inspiratory noise; the momentary airway geometry filters and modifies this noise (i.e., resonances in the airway acting as band-pass filters to the noise, Footnote 17). That is, this postapneic snore consists of irregular high frequency *noise* with poorly formed and variable *bursts* in the time domain (Perez-Padilla et al. 1993). In the frequency domain, this snore shows a broad spectral peak at around

[21] Namely, *obstructive snoring* is considered as a primary symptom for *sleep apnea* (Brunt et al. 1997). However, the noisy respiration during sleep, as actually the snoring corresponds to, can not be used as a sole indicator of breathing abnormalities, such as sleep apnea (Wilson et al. 1999). Likewise, snoring lacks *specificity* for diagnosis of apneas.

450 Hz and another one at around 1 kHz; both peaks probably raised by filtering and modification of the noise in the airway. The latter bursts of sound—superimposed on the noise—also indicate additional sound sources other than air turbulences (*turbulent mechanisms*), such as intermittent opening and closing of the airway (*vocal mechanisms*), all sources contributing to the first postapneic snore.

In the course of *subsequent breaths* after the first postapneic snore, apneic sounds become like obstructive snoring sounds. The *high-pitched sounds* may possibly fade into the *low-pitched sounds* in the ventilatory interval between two neighbouring apneas. In fact, postapneic sounds exhibit very high variability in the time and frequency domains because of highly pronounced morphological variations of the pharyngeal airway from fully occluded to fully dilated (Perez-Padilla et al. 1993).

Figure 4.15 illustrates *apneic sounds* and their variability considering different types of apneas. The *obstructive sleep apnea* (Fig. 4.15a) is surrounded by obstructive snoring events because this type of apnea is characterised by obstructive occlusions of the upper airway. In analogy to Fig. 4.14b, high frequency components above 800 Hz can be clearly distinguished, which is a typical feature of obstructive snoring (Sect. 4.1.1.3). Brief high-pitched sounds can be observed in the middle of this obstructive apnea, which indicate *apneic respiratory efforts* during the apnea. These acoustically noticeable efforts denote time intervals during which the airway is incompletely occluded. Distinct snoring surrounds the depicted *central sleep apnea* (Fig. 4.15b), which composition in the time and frequency domains is similar to that of loud snoring from Fig. 4.13b. In the case of the *mixed sleep apnea* (Fig. 4.15c), a preceding central segment without apneic respiratory efforts can be seen, followed by an obstructive segment with numerous apneic respiratory efforts; in fact, this observation complies with the definition of the mixed sleep apnea (Sect. 3.1.2).

The acoustical fingerprint of an *obstructive sleep hypopnea*—characterised by a mere reduction of the respiratory airflow (Sect. 3.1.2)—is demonstrated in Fig. 4.16. Temporal reduction of the amplitude (or the intensity) of the obstructive snoring sounds can be observed in the time domain; compare the depicted envelopes in Fig. 4.16a. In the spectrogram, the power of high frequency components above around 800 Hz decreases temporarily during the hypopnea (Fig. 4.16b). However, snoring events during the hypopnea still remain as obstructive snoring events because of the obvious spectral gap around 800 Hz in the obstructive events; compare with Fig. 4.14b.

4.1.1.5 Mutual Interrelations

As shown in Sects. 4.1.1.1–4.1.1.3, *heart sounds* arise in the course of cardiac activity while *lung, snoring, and apneic sounds* arise in the course of respiratory activity. Since *cardiac and respiratory activities* are intimately and conclusively related to each other (Sect. 3), body sounds originating from cardiac and respiratory activities (Fig. 4.1) exhibit various *mutual interrelations*. Likewise, sources of the

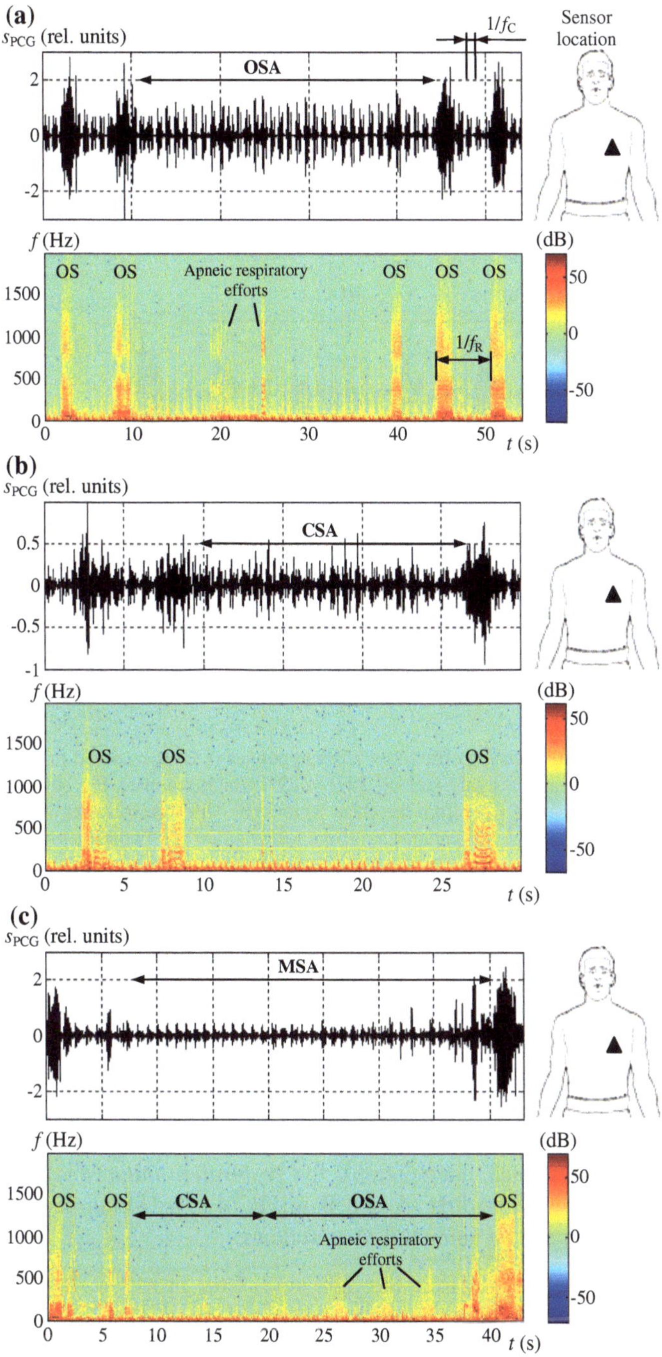
(a)
s_{PCG} (rel. units)
$1/f_C$
Sensor location
OSA
f (Hz)
(dB)
OS
OS
Apneic respiratory efforts
OS
OS
OS
$1/f_R$
t (s)
(b)
s_{PCG} (rel. units)
CSA
f (Hz)
(dB)
OS
OS
OS
t (s)
(c)
s_{PCG} (rel. units)
MSA
f (Hz)
(dB)
OS
OS
CSA
OSA
OS
Apneic respiratory efforts
t (s)

◀**Fig. 4.15** Apneic sounds during sleep, as illustrated by an acoustic biosignal phonocardiogram s_{PCG} from the chest region of sleep apnea patients. (**a**) Obstructive sleep apnea (OSA) with an apneic respiratory effort, surrounded by obstructive snoring (OS) events (male patient, BMI = 29 kg/m^2); compare Fig. 4.14. (**b**) Central sleep apnea (CSA) delimited by OS events (female, BMI = 28.2 kg/m^2). (**c**) Mixed sleep apnea (MSA) with successive segments of CSA and OSA (male, BMI = 35.5 kg/m^2). The heart rate f_C and respiratory rate f_R are indicated. The corresponding spectrograms (*lower subfigures*) disclose intermittent respiratory activity in more detail. For parameters of the spectrograms see Footnote 4

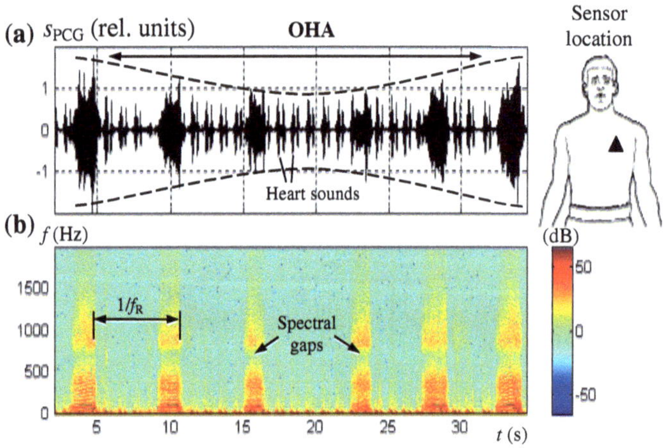

Fig. 4.16 Apneic sounds during obstructive sleep hypopnea (OHA), recorded from a patient with the obstructive sleep apnea (male patient, BMI = 29 kg/m^2); compare Fig. 4.15. (**a**) Acoustic biosignal phonocardiogram s_{PCG} from the chest region. (**b**) The corresponding spectrogram with indicated respiratory rate f_R. For parameters of the spectrogram see Footnote 4

corresponding body sounds—in terms of the sound's formation in the electrical circuit model, see Fig. 4.2—are tightly related to each other. On the other hand, mutual interrelations among body sounds originating from *respiratory activity* only, such as lung, snoring, and apneic sounds, also exist because all these sounds have the same origin—the respiration.

Figure 4.17 illustrates the latter relationships of sources of the different body sounds. In fact, *mechanic*, *neurogenic*, and *hormonal* control mechanisms are involved here, as described below (compare Sect. 3.2).

To begin with mutual interrelations, the *respiration-induced effects on heart sounds* will be considered first (A and B in Fig. 4.17). During *inspiration* these modulation effects can be summarized as follows:

- intensification of sounds from the right side of the heart, i.e., intensification of *right-sided heart sounds* which are generated by closure of the right-sided valves, the tricuspid and pulmonary valve (Fig. 4.3);
- attenuation of *left-sided heart sounds*, generated by closure of the left-sided valves, the mitral and aortic valve (Fig. 4.3);

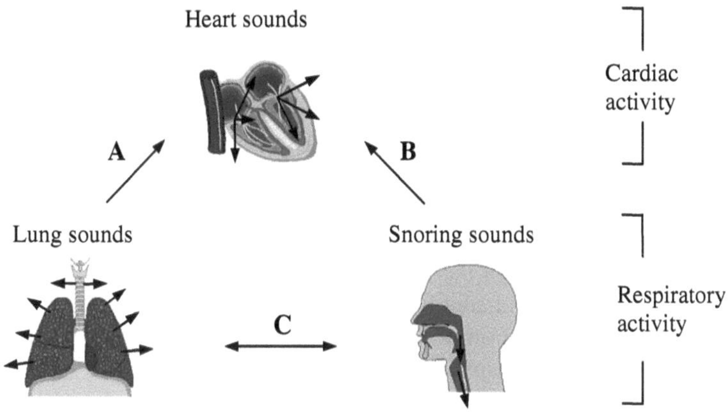

Fig. 4.17 Mutual interrelations of the sources of the different body sounds with indicated direction of the physiological influence

- (intensified) *splitting* of the first and second heart sound; and
- increased *repetition rate* of heart sounds, i.e., increased f_C.

Obviously, modulation effects reverse during expiration and disappear when *holding breath*. Generally, the changing volume of the lung influences the pressure conditions within the heart and those close to the heart, which in turn mechanically influences *intensity and timing* of the *valve's closure*.

As described in section "Normal Respiration" in Sect. 3.2.1.2 in more detail and illustrated in Fig. 3.31, during *inspiration* the *right ventricular stroke volume* increases temporarily and rises the volume of the *decelerated blood* in the right side of the heart at the valve's closure. Thus, the vibration intensities of the corresponding *right-sided* valves (Fig. 4.3) and the involved right-sided blood volumes increase, which intensifies the *right-sided heart sounds*; compare generation mechanisms of heart sounds in Sect. 4.1.1.1. Conversely, during inspiration the left ventricular stroke volume decreases temporarily, which causes the *left-sided heart sounds* to decrease in their intensity.

Likewise, authors in Amit et al. (2009) report that the decreased left ventricular stroke volume and the correspondingly decreased left ventricular *contraction force* (compare Footnote 225 in Sect. 3) contribute to the attenuation of the first heart sound during inspiration. In analogy, an increased *pressure difference* between aortic pressure and left ventricular pressure (i.e., increased *afterload*) accentuates the second heart sound. Therefore, *mechanical mechanisms* are responsible for the rhythmic changes in the sound intensity within the respiration cycle.

Similar *mechanical mechanisms* are responsible for an audible separation between consecutive sound components within the first heart sound and those within the second heart sound (Sect. 4.1.1.1), i.e., responsible for an intensified *splitting* of these heart sounds during *inspiration*. As described in section "Normal Respiration" in Sect. 3.2.1.2 and illustrated in Fig. 3.32, the tricuspid valve and pulmonary valve (Fig. 4.3) stay open longer in the course of the ventricular systole during inspiration in comparison with expiration; this is because the right ventricular stroke volume increases during inspiration. In addition, the mitral valve closes a bit earlier than the tricuspid valve due to mechanisms described in Footnote 2. The aortic valve closes earlier than the pulmonary valve because of both the decreased left ventricular stroke volume and increased right ventricular stroke volume. As a result, the gap between the early sound contribution from the closure of the mitral valve and the late sound contribution from the closure of the tricuspid valve widens within the *first heart sound* at inspiration. In analogy, the gap between the early sound contribution from the closure of the aortic valve and the late sound contribution from the closure of the pulmonary valve widens within the *second heart sound* at inspiration.

In addition to the discussed mechanisms, the *first and second heart sound* was observed to be slightly *delayed and advanced*, respectively, during inspiration (Amit et al. 2009). The latter observation corresponds to the dominance of the right-sided sounds in the first heart sound and the dominance of the left-sided sounds in the second heart sound, given the discussed changes in stroke volumes; compare Fig. 4.18a.

Neurogenic mechanisms account for *increased* f_C at *inspiration*, whereas f_C decreases at expiration. As described and illustrated in Sect. 3.2.1.1, respiratory sinus arrhythmia occurs from the influence of breathing on the autonomic nervous system which governs the heart beat (Sect. 3.1.1).

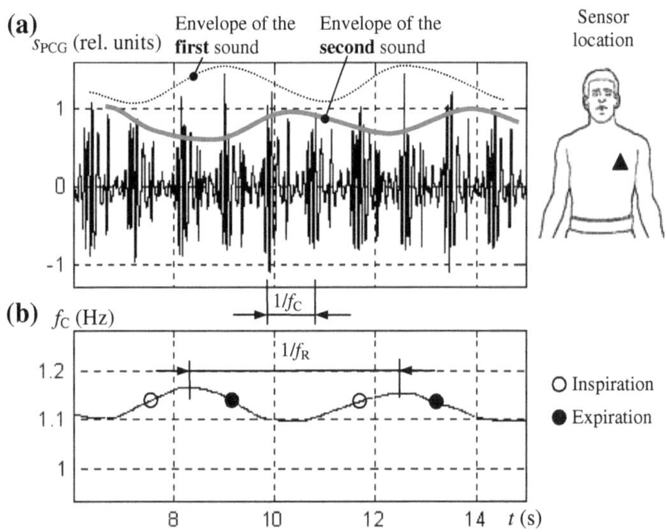

Fig. 4.18 Influence of breathing on heart sounds. (**a**) Acoustic biosignal phonocardiogram s_{PCG} from the chest region with an added envelope which indicates the amplitude modulation of s_{PCG} with the respiratory rate f_R. (**b**) The instantaneous heart rate f_C derived from s_{PCG}

In fact, the aforementioned mechanisms rather apply for *normal breathing* (A in Fig. 4.17). However, in the case of *obstructive snoring* the long-term impact of the snoring sound sources on the heart sound sources becomes more intricate and less direct (B in Fig. 4.17). The persistent obstruction of the upper airways may overload the heart in the long-term, involving also hormonal mechanisms, favouring *cardiovascular diseases* (Sect. 4.1.1.3) and thus causing malfunction of valves and vibrating structures (i.e., sound sources) in and close to the heart. In particular, impact on the heart is strong if the obstruction occurs with an intermittent and complete closure of the airway lumen, i.e., with intermittent apneas (Sect. 3.1.2).

Body sounds of the respiratory origin arise in synchrony, yielding an identical respiratory rate (C in Fig. 4.17). Nonetheless, signal properties of lung, snoring, and apneic sounds remain very different; see Sects. 4.1.1.2–4.1.1.4. In addition, it can be expected that laboured breathing or heavy obstructive snoring (with an intermittent occlusion of the upper airways) could temporarily alter resonance (spatial) characteristics of airways in which lung sounds originate. This would lead to snoring-related changes in spectral components of lung sounds; compare Footnote 17 and section "Specific Issues" in Sect. 4.1.2.1.

Figure 4.18 illustrates *respiration-induced effects on heart sounds*. It can be observed that the amplitude of the *first heart sound* increases during *inspiration* while that of the *second heart sound* decreases, see the corresponding envelopes in Fig. 4.18a. That is, the amplification of the right-sided heart sounds is stronger in the first heart sound than the concurrent attenuation of the left-sided heart sounds, i.e., the right-sided sounds dominate in the first heart sound. The reverse is true for the second heart sound. However, as noticed in section "Normal Respiration" in Sect. 3.2.1.2, the illustrated behaviour in Fig. 4.18a *is not generally valid*, as also reported in Amit et al. (2009). For instance, Fig. 3.31b shows decreasing amplitude of the first heart sound and increasing amplitude of the second heart sound during inspiration; Figs. 4.9a and 3.5b demonstrate increasing both first and second heart sounds during inspiration; Fig. 3.32b discloses decreased intensities of both heart sounds during inspiration, as also reported in Ishikawa and Tamura (1979). In addition, Fig. 4.18b illustrates a temporal *increase of* f_C (i.e., the reoccurrence rate of heart sounds) at *inspiration*, as expected from respiratory sinus arrhythmia.

Lastly, *overlapping frequency ranges* of the different *body sounds* should be addressed. As can be extracted from Sects. 4.1.1.1–4.1.1.3,

- *heart sounds* reside in the approximate frequency range up to 100 Hz,
- tracheobronchial *lung sounds* in the range of 100–1,000 Hz,
- vesicular *lung sounds* 100–500 Hz,
- normal *snoring sounds* 100–800 Hz, and
- obstructive *snoring sounds* 100–2,000 Hz.

Besides the above approximations of the frequency ranges, the frequency components of heart sounds overlap with that of breathing sounds (such as lung sounds and snoring sounds). In particular, the relatively *strong heart sounds* overlap with the low frequency components of the relatively *weak breathing sounds*. For instance, the *interference* of *heart sounds* in *breathing sounds*—as recorded on the neck—was quantitatively reported in Lessard and Jones (1988). The authors showed that the contribution of heart sounds can not be neglected even at frequencies above 100 Hz because of their relatively high intensity (Sect. 4.1.1.2). The first heart sound was shown to contribute to the acoustic power in the frequency range of 175–225 Hz during inspiration and 75–125 Hz during expiration. In parallel, the second heart sound—with spectral components of even higher frequencies (Sect. 4.1.1.1)—appeared to contribute to the acoustic power in a more extended range of 75–425 Hz during inspiration and 75–325 Hz during expiration.

4.1.2 Transmission of Body Sounds

The *transmission of body sounds* throughout the tissue, in addition to their genesis (Sect. 4.1.1), comprises *formation aspects* of body sounds, according to the model of acoustic biosignals (Fig. 4.2). As illustrated in Fig. 4.1, the acoustical path of body sounds begins within the respective *sound source* which is given by oscillating (biological) structures, vibrating blood volumes and turbulent air. The induced *mechanical waves*[22] *propagate* through the tissue along multiple paths and are subjected to changes in their intensity (mostly damping) because of *absorption*, *scattering*, *diffraction*, *reflection*, *refraction*, and *resonance* phenomena. In fact, a large percentage of the sound energy dissipates on the way and never reaches the skin surface where an acoustic *sensing device* is usually located (Fig. 4.1).

---

[22] *Sound* is provided by *mechanical oscillations* in an *elastic medium*, as illustrated in Fig. 4.19a for longitudinal waves. Under influence of a transient external force, composing *particles* of the medium (e.g., molecules in the air or tissue) are dislocated from their equilibrium (rest) position and are then left to their own devices. Inertial and elastic forces (restoring forces) are induced, which force these particles to move back, so that the particles start to swing around their equilibrium positions with a certain *particle velocity* in terms of mechanical oscillations. Consequently, as demonstrated in Fig. 4.19, the mechanical *overpressure* (positive *sound pressure*) arises in the regions of increased medium density while the *underpressure* (negative sound pressure) arises in the regions of decreased density, as compared with the resting state of the medium without propagating sounds; compare Footnote 26. Please note that

- the spatial *wave of the sound pressure* $p(x)$ propagating in the direction x (Fig. 4.19b) is in-phase with the *wave of the particle velocity* $u(x)$ in *unlimited elastic medium* (but not in the limited resonating cavity such as in Fig. 4.24), whereas
- the corresponding *wave of the particle deflection* is dislocated by 90° with respect to $p(x)$ or $u(x)$.

In fact, the *particle velocity* strongly differs from the *sound propagation velocity* (4.3). The particle velocity (4.6) is usually by many orders lower than the propagation velocity; e.g., $5 \cdot 10^{-8}$ m/s versus 343 m/s in the air (Table 4.1) at the human auditory threshold at 1 kHz (Veit 1996). In addition, the particle velocity increases with increasing loudness (sound intensity) while the propagation velocity usually does not.

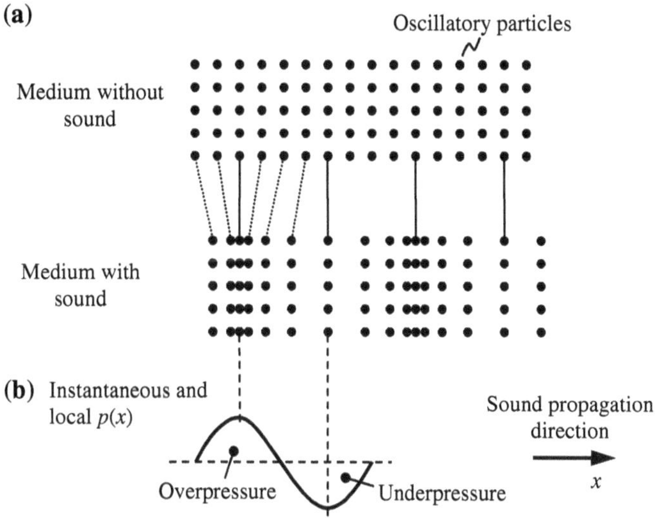

Fig. 4.19 Sound waves or mechanical deformation propagating in an elastic medium. (**a**) Oscillatory particles in medium—elastically bound to each other—are transiently dislocated from their equilibrium position by the propagating pressure wave. (**b**) The associated sound pressure p indicates local regions of overpressure and underpressure. In fact, a longitudinal density wave is demonstrated here, while the wave p oscillates in the direction of the sound propagation

4.1.2.1 Propagation of Sounds

General Issues

Body *sounds propagate* in a *biological medium* with the *sound propagation velocity* v (a *time-space* characteristic), oscillate with the sound frequency f in the time domain (a *time* characteristic), and oscillate with the wavelength λ along their propagation path (a *space* characteristic) according to

$$v = \lambda \cdot f\ . \tag{4.3}$$

In fact, the above equation applies for any type of wave propagation, including the propagation of pulse waves along arteries, as discussed in section "Pulse Propagation" in Sect. 2.5.2.3.

The value of v is determined by (macroscopic) physical properties of the propagation medium, such as biological tissue or air, according to

$$v = \sqrt{\frac{\kappa}{\rho}} = \sqrt{\frac{1}{\rho \cdot D}}\ . \tag{4.4}$$

Table 4.1 Typical and approximate sound velocities v in the air, water, muscle, bone (Veit 1996), large airways (i.e., with diameter > 1 mm), biological tissue (Kompis et al. 2001), tallow (Trendelenburg 1961), and lung tissue (Kompis et al. 2001; Rice 1983; Wodicka et al. 1989)

Propagation medium	Sound velocity v (m/s)	Sound wavelength λ (m)	Absorption coefficient $\alpha_F + \alpha_T$ (1/m)
Air	343	0.34	10^{-5}
Water	1,440	1.44	10^{-8}
Sea water	1,533	1.53	10^{-8}
Fat	1,450	1.45	$> 10^{-8}$
(Olive) oil	1,420	1.42	$> 10^{-6}$
Tallow	390	0.39	10^{-4}
Large airways	270	0.27	10^{-5}
Tissue	1,500	1.5	$> 10^{-8}$
Muscle	1,560	1.56	$> 10^{-8}$
Bone	3,600	3.6	$> 10^{-4}$
Lung tissue	50	0.05	$> 10^{-4}$

The corresponding wavelengths λ are estimated for the sound frequency 1 kHz by (4.3). Typical and approximate sound absorption coefficients α (considering only α_F and α_T, see (4.14)) are also given for the frequency 1 kHz, according to the classical sound absorption theory (Meyer and Neumann 1975; Trendelenburg 1961). Data has been accumulated from different sources (Kaniusas 2007)

In analogy to 2.22 and 2.23, κ is the module of volume elasticity, ρ the density, and D (= $1/\kappa$) the compliance (or adiabatic compressibility) of the propagation medium.

Table 4.1 summarizes v *and* λ for the most relevant types of *physical and biological media* involved in the transmission of body sounds. For the sake of simplicity, only approximate values are given without considering effects of varying f, temperature, humidity, and the type of acoustical waves (e.g., longitudinal or transverse wave).[23]

[23] The *influence of temperature and humidity* on v—and thus also on λ (4.3)—should be discussed shortly from a physiological point of view. It is well known that v in the air tends to increase with increasing *temperature*, yielding an increase rate of about 0.6 m/s per degree Celsius. During *inspiration* the air at room temperature (usually < 37 °C) enters the respiratory airways, whereas during *expiration* the warmed up air at body temperature (≈ 37 °C) leaves the airways. Consequently, the level of v in the large airways decreases with inspiration by a few percent and correspondingly increases with expiration.

Regarding the influence of *humidity*, it should be noted that the inspired air is saturated with water vapour (relative humidity of 100 %) as it flows over the wet and warm mucous membranes lining the respiratory airways (Sect. 2.6.2). The effective value of v is very slightly influenced by the air humidity, e.g., a change in the relative humidity from 50 % at *inspiration* to 100 % at *expiration* increases v by only about 0.5 % at 37 °C.

In fact, the more *compressible* is the propagation medium, i.e., the larger is D (4.4), the lower is v. Consecutively, the air, large airways, and lung tissue exhibit relatively low values of v, whereas incompressible liquids like (sea) water show relatively high values of v. Solid substances such as *bone* tend to show even higher values of v. *Biological tissues* (including blood) yield v comparable with that in the (sea) *water* because tissues have a relatively high (salt) water content of about 60 % (Silbernagl 2007); compare Table 2.1. The *lung tissue*, given by a mixture of a compliant tissue and the air,[24] yields the lowest v in the order of 50 m/s, or in the range of 23–60 m/s (Kompis et al. 2001). It can also be observed in Table 4.1 that fat (e.g., fat layers in tissue) tends to slow down the sound propagation; for instance, the level of v in porcine muscle and skin (1,620 and 1,680 m/s) were reported as being higher than in the outer (skin) fat layer (1,435 m/s) (Koch et al. 2010).

Specific Issues

The calculated *values of* λ for $f = 1$ kHz (Table 4.1) indicate that they are in the range of average *body dimensions* of less than 2 m but are significantly larger than the *auscultation distance* r from sources of inner body sounds to typical auscultation sites. Figure 4.20 demonstrates that typical values of r are in the range of 5–30 cm. That is, the acoustical *near field*—satisfying the inequality $r < 2 \cdot \lambda$ (see Sect. 6)—dominates in typical *auscultation sites* on the skin. In particular, this near field condition is fulfilled in *solid* and *liquid* media (with relatively high κ, (4.4)) while in the *air* r and λ become comparable in size. However, the *lung tissue* is an exception with λ in the

---

[24] The value of v *in the lung tissue* depends strongly on the air content in the lung. Provided that the volumetric portion of the air is 75 % and the rest is tissue (Wodicka et al. 1989), the effective ρ and D of the composite mixture can be estimated as

$$\rho = 0.75 \cdot \rho_{\mathrm{A}} + 0.25 \cdot \rho_{\mathrm{T}} \approx 0.25 \cdot \rho_{\mathrm{T}}$$

and

$$D = 0.75 \cdot D_{\mathrm{A}} + 0.25 \cdot D_{\mathrm{T}} \approx 0.75 \cdot D_{\mathrm{A}}$$

where ρ_{A} (= 1.2 kg/m^3) and ρ_{T} (= 1,040 kg/m^3) are approximate densities of the air and tissue, respectively. In analogy, D_{A} (= 7,083 GPa^{-1}) and D_{T} (= 0.43 GPa^{-1}) are the corresponding compliances which are estimated using (4.4) with v (Table 4.1) and ρ as parameters. It can be observed that $\rho_{\mathrm{A}} \ll \rho_{\mathrm{T}}$ and $D_{\mathrm{A}} \gg D_{\mathrm{T}}$. With the effective ρ and D from above equations, (4.4) yields v = 27 m/s fitting well the reported range of 23–60 m/s (Kompis et al. 2001).

In fact, the above postulation of a *homogenous mixture of gas and tissue* assumes that the *size of* λ in the lung parenchyma is significantly larger than the alveolar size (diameter < 1 mm). In fact, this assumption is entirely met by body sounds in the frequency range up to 2 kHz (see section "Volume Effects" in Sect. 4.1.2.2).

range of only 5 cm (at $f = 1$ kHz). Thus, the condition of the near field is hardly met in the lung tissue (and the far field conditions apply).

It should be stressed that the size of λ *decreases* with *increasing* f (4.3), so that the inequality $r < 2 \cdot \lambda$ increasingly ceases to apply. In addition, as will be shown later (in Fig. 4.23), high frequency components of body sounds tend to take an airway bound route while propagating in the body; i.e., the branched structure of the (air-filled) respiratory airways is preferred over the lung parenchyma (or semi-solid tissue of the inner mediastinum) as the *propagation pathway* of body sounds.

In the case of *body sounds*, two types of their *sources* can be distinguished:

- *point source* of sound, i.e., the spatial extension of a single sound source is less than r and is limited to a particular region of the body;
- *diffuse source* of sound, i.e., distributed multiple sound sources dominate within a relatively large region of the body, which dimensions are in the range of r or even exceed r.

For instance, sound sources of *heart sounds*, tracheobronchial *lung sounds*, and *snoring sounds* can be approximated as point sources of sound (compare Sect. 4.2.2.3). In contrast, vesicular *lung sounds* yield a diffuse source of sound.

The type of sound source is relevant for a *qualitative understanding* of the propagation and absorption of body sounds, whereas the *sound attenuation* over a propagation distance is basically governed by both

- propagation *geometry* and
- propagation *medium*.

In the case of the *point source* of sound and thus *spherical waves* (Sect. 6), the sound intensity I at the distance r obeys the *geometry-related damping*, namely, the *inverse square law*[25]; here free and far field radiation is assumed, i.e., without sound absorption in the propagation medium and without sound reflections at any limiting boundary surfaces. Since I is inversely related to r^2 (Fig. 4.21) and

$$I = p \cdot u = \frac{p^2}{Z} \tag{4.5}$$

[25] The *inverse square law* applies for *spherical waves* (Sect. 6) when *sounds* are radiated in *lossless media* outward radially from a point source, as illustrated in Fig. 4.21. Since the original *source power* P is spread out over an *area* ($= 4\pi \cdot r^2$) of a sphere, which increases in proportion to r^2 with the velocity v, the resulting *sound intensity* I at the distance r (passing through a unit area and facing directly the point source) is equal to

$$I = \frac{P}{4\pi \cdot r^2},$$

i.e., is inversely related to r^2. As demonstrated in Fig. 4.21, the level of I quadruples while p doubles when r is halved.

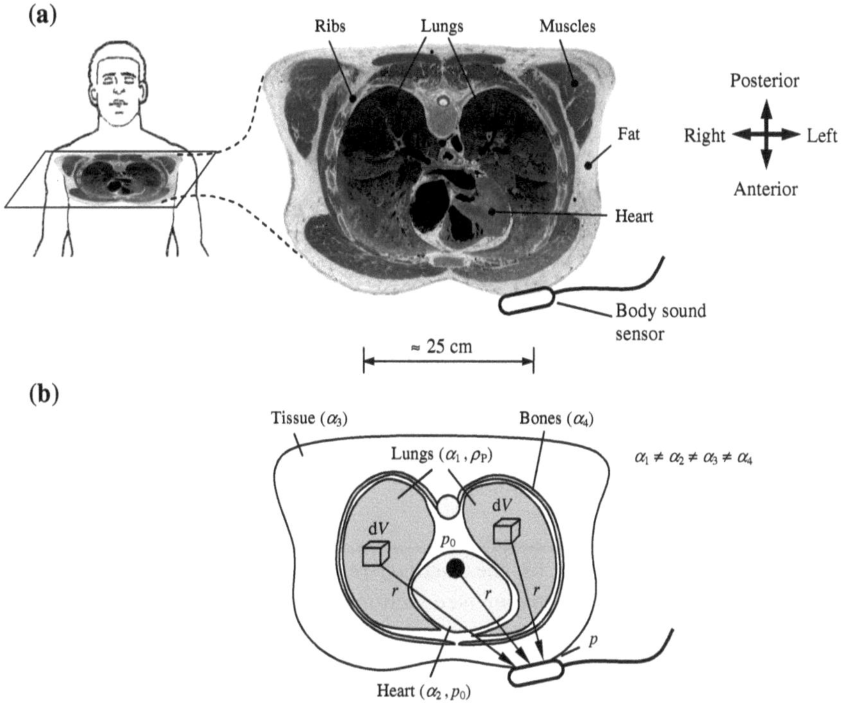

Fig. 4.20 Propagation of body sounds in the thorax. (**a**) Photographic image of the cross-section of the thorax at the level of the heart (Bulling 1997), disclosing a highly heterogeneous propagation medium. (**b**) Schematic representation of the cross-section of the thorax at the level of the heart. Contributions of a point source of heart sounds (with the induced sound pressure p_0 within the source, (4.7)) and diffuse (distributed) sources of lung sounds (with the volume density ρ_P of the induced sound pressure within the differential volume dV, (4.8)) to the sound pressure p are indicated at an auscultation site on the chest where an acoustic sensing device resides. The resulting level of p depends strongly on the source-sensor distance r and sound absorption coefficients α

with Z as the *characteristic acoustic impedance* (compare analogous concepts in Sect. 6), which is a (macroscopic) material property given by

$$Z = \frac{p}{u} = \rho \cdot v, \tag{4.6}$$

it can easily be derived that the effective magnitude of the *sound pressure p is inversely related to r* (Fig. 4.21) in a *spherical wave*. Here u denotes the effective magnitude of the *particle velocity* (around the particle's equilibrium), see Footnote 22. However, (4.5) strictly applies only under conditions of *far field* ($r > 2 \cdot \lambda$) where the spatial fields of p and u oscillate *in-phase*; compare Sect. 6.

In addition to the aforementioned geometry-related damping, the propagation medium absorbs sounds, i.e., the absorbing medium reduces I with increasing r,

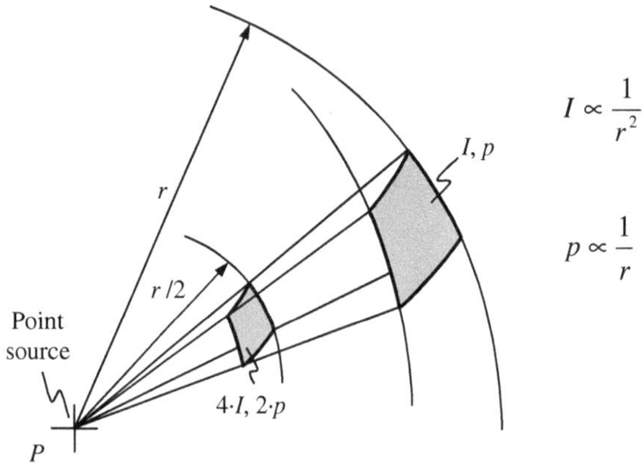

Fig. 4.21 Illustration of the inverse square law for spherical waves with r as the distance from the point source to the auscultation site, P the source power (at $r = 0$), and I the resulting sound intensity at the distance r; compare Sect. 6

which will be referred to as the *medium-related damping*. Because of this absorption, the *medium is heated up* (see section "Volume Effects" in Sect. 4.1.2.2). In particular, the *envelope of the sound pressure p*[26] experiences *exponential decay* in a *homogenous medium* (Sect. 6), which rate increases with the *sound absorption coefficient* α. Likewise, the local decrease of p is proportional to the amplitude of p itself; see Footnote 19 in Sect. 5. Figure 4.22 demonstrates the spatial relationship between the envelope of p, the instantaneous p, and r considering only the medium-related damping. The medium with a higher α (= α_1 and $\alpha_1 > \alpha_2$) yields a steeper decrease of the envelope of p and thus a stronger damping of body sounds within this medium; see the corresponding tangents in Fig. 4.22.

Considering both mechanisms of the *sound attenuation* with r as parameter, the effective level of p at the distance r—provided that a *point source of sound* is given yielding *spherical waves*—can be approximated as

$$p = c \cdot \frac{p_0}{r} \cdot e^{-\alpha \cdot r}. \tag{4.7}$$

Here p_0 describes the induced sound pressure within the point source (at $r = 0$) and c is a positive constant; compare Fig. 4.20b. The factor $1/r$ accounts for the geometry-related damping while the factor $e^{-\alpha \cdot r}$ accounts for the medium-related damping. It is interesting to observe in (4.7) that the rate, with which p decays,

[26] It should be stressed that the *sound pressure p* is an *overpressure* (and the corresponding *underpressure*) related to the ambient atmospheric pressure; compare Fig. 4.19. Consequently, positive or negative p means pressure above or below the ambient pressure, respectively. In this context, Fig. 4.22 demonstrates the decay of this overpressure (or the decay of the underpressure), whereas Fig. 4.24 demonstrates periodic changes of the *instantaneous sound pressure* from values below the ambient pressure to that above the ambient pressure and vice versa.

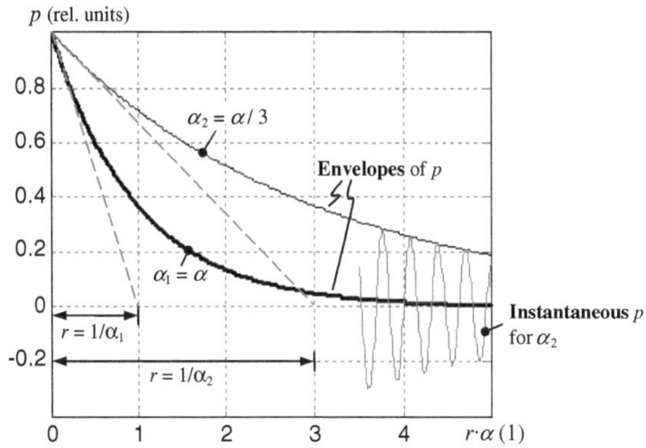

Fig. 4.22 Spatial response of the envelopes of the sound pressure p (Footnote 26) in terms of the medium-related damping as a function of the propagation distance r at two different sound absorption coefficients α_1 and α_2 ($< \alpha_1$). The corresponding tangents are depicted at $r = 0$ (*dashed lines*) to indicate the size of α (compare Sect. 6)

decreases with increasing r. Likewise, *spherical waves* close to the point source mutate into *plain waves* distant to the source.

In analogy with (4.7), a *diffuse source* of sound yields

$$p = c \cdot \int_V \frac{\rho_P}{r} \cdot e^{-\alpha \cdot r} \cdot dV \ , \tag{4.8}$$

where ρ_P is the volume *density of the induced sound pressure* within the differential volume dV located in the diffuse source. The induced local (and differential) sound pressure dp_O in the diffuse source could be given as $\rho_P \cdot dV$, without considering pressure contributions from neighbouring regions. In general, the density ρ_P depends on r, i.e., on the spatial location of the volume dV.

Provided that a diffuse source of sound is given, it can be expected that the geometry-related damping dominates less than around a point source because the sound wavefront of the diffuse source resembles rather a *plane wave*. In consequence, the geometry-related factor should be less strong than $1/r$.[27]

[27] Generally, different *assumptions* regarding the *geometry-related damping factor*, i.e., the factor $1/r$ from (4.7), can be found in literature. For instance, this factor was completely neglected in Wodicka et al. (1989), assuming plain wave conditions for the propagation of the intensity I in the lung parenchyma ($I \propto p^2$, (4.5)). In contrast, authors in Kompis et al. (1998, 2001) assumed an even stronger damping factor $1/r^2$ for the assessment of the spatial distribution of the effective p in the thorax.

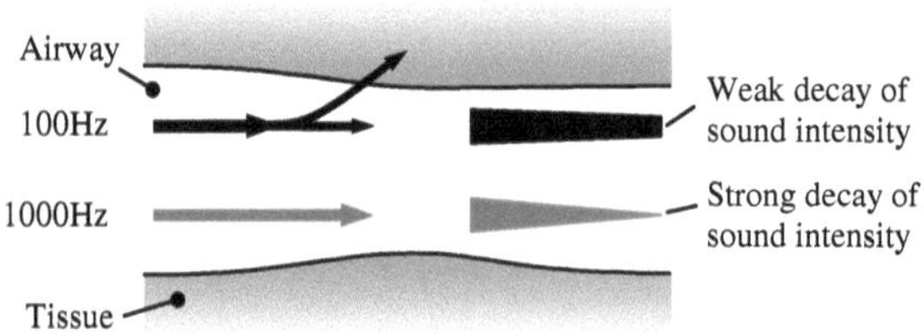

Fig. 4.23 Sound propagation and attenuation as a function of the sound's frequency; compare Fig. 4.1

Figure 4.20 depicts photographically and schematically sources of body sounds and their respective distances r to an auscultation site on the chest wall. The geometrical dimensions of the thoracic cross-section, in combination with data from Table 4.1, prove the discussed inequality $r < 2 \cdot \lambda$. Figure 4.20b demonstrates also the integration procedure related to (4.8) for the highly heterogeneous medium in the thorax (Fig. 4.20a), whereas each medium has its own α (Table 4.1). Provided that a point source and a diffuse source are simultaneously active, the resulting p at the auscultation site shows *additive contributions* from this point source with p_0 (e.g., located in the heart) and this diffuse source with $\mathrm{d}p_0 = \rho_\mathrm{P} \cdot \mathrm{d}V$ (e.g., located in the lung).

It is interesting to observe that the *propagation pathway* of body sounds differs with the varying *sound frequency*. In particular, it applies for *lung sounds* and *snoring sounds* propagating in the highly heterogeneous medium which is basically composed of waterlike tissue and the air:

- *Low frequency sounds*, i.e., below 300 Hz (Pasterkamp et al. 1997b) or in the range of 100–600 Hz (Wodicka et al. 1989), tend to be coupled from the respiratory airways (including large airways such as trachea) into the surrounding lung parenchyma or inner *mediastinum* via induced mechanical *oscillations of airway walls*. The network of airway branches behaves as a network of *compliant tubes* which *non-rigid walls resonate* in response to the intraluminal sound at these relatively low frequencies; i.e., airway walls tend to absorb energy of low frequency sounds. Likewise, the travelling of low frequency sounds is impeded down the air-filled airways, whereas sounds bypass other airways.
- *High frequency sounds*, i.e., with frequencies above those of low frequency sounds (see above), experience *rigid walls* of airways because of their inherent mass. In other words, walls are too inert to follow fast mechanical vibrations of sounds so that these sounds remain mainly within the airway lumen and travel further along the network of *airway branches*.

In other words, *heart sounds*, i.e., low frequency sounds, tend to remain within the *mediastinum*. As illustrated in Fig. 4.23, *low frequency sounds* are primarily bound to the *lung parenchyma* or *inner mediastinum*, whereas these sounds exit respiratory airways. According to Rice (1983), translobar sounds tend to travel through the bulk of the lung parenchyma—a foamlike substance, a homogenous mixture of waterlike tissue and air (Footnote 24)—and not along airways or blood vessels because the parenchyma acts as an elastic continuum to audible sounds. In contrast, the propagation of *high frequency sounds* is primarily linked to *airway-bound routes* (Fig. 4.23).

Such observations suggest that *low frequency sounds* at the *chest wall* provide information mostly on the lung parenchyma and inner mediastinum, whereas *high frequency sounds* at the chest wall reflect mostly airway properties. This dependence of the *propagation pathway* on the sound frequency has *strong implications* on the composition of sounds registered by a room microphone or skin microphone (Sect. 4.2.1.2) and on the asymmetry of the sound transmission in the thorax (Sect. 4.2.2.3).

Provided that the propagation pathway of body sounds varies with the sound frequency, it can be deduced from Table 4.1 that the *sound propagation velocity* v will also *depend on the frequency*. For instance, low frequency components of lung sounds will propagate with lower v than high frequency components. This is because these low frequency components are mainly bound to the parenchymal tissue with $v \approx 50$ m/s while the high frequency components are mainly bound to airways with $v \approx 270$ m/s.[28]

The sound propagation in *spatially limited air volumes* such as the upper airways should be addressed in more detail. Provided that the axial dimensions of the airways are in the range of λ of a propagating sound wave (or even larger than λ), which is easily met especially for higher sound frequencies (Table 4.1 and (4.3)), the upper airways act as *resonating acoustic filters*. Such filters attenuate the transfer of sound energy at certain frequencies while allowing maximal energy through at particular resonance frequencies, also known as *formant frequencies*.

[28] Various *experimental data* confirm the dependence of the propagation pathway on the sound frequency and thus the *dependence of* v *on the frequency*. The authors in Pasterkamp et al. (1997b) demonstrate that low frequency sounds at 200 Hz are transmitted from the trachea to the chest wall with a *phase delay* of about 2.5 ms, whereas high frequency sounds at 800 Hz traverse a faster route with a phase delay of only 1.5 ms. For an assumed propagation distance of 20 cm, it would yield $v \approx 80$ m/s for low frequency sounds and $v \approx 130$ m/s for high frequency sounds.

The hypothesis of parenchymal propagation of sounds at lower frequencies is also supported by the fact that the inhalation of a *helium-oxygen mixture* (80 % helium and 20 % oxygen) affects only weakly (i.e., reduces) the *phase delay* of the sound transmission from the trachea to the chest wall at *lower frequencies*, in comparison with the inhalation of air (Pasterkamp et al. 1997b). In contrast, this *phase delay* is significantly reduced at *higher frequencies* while inhaling the helium-oxygen mixture. In quantitative terms, a reduction by about 0.7 ms was observed at 800 Hz (i.e., from 1.5 ms for the air inhalation down to 0.8 ms for the gas mixture) with almost no reduction at 200 Hz (i.e., 2.5 ms for both the air and gas mixture). Since the helium-oxygen mixture shows higher value of v than the air, the discussed observation proves a predominantly airway-bound sound transmission of high frequency sounds in the thorax.

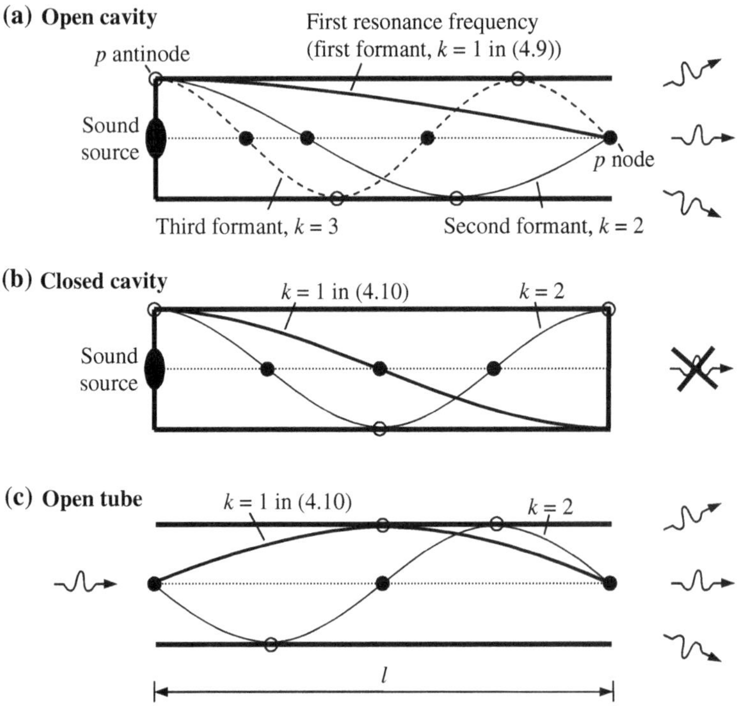

Fig. 4.24 Resonating cavities and openings with the corresponding waveforms of the sound pressure p. (**a**) Resonance of the open cavity, e.g., the upper airway which is approximated as a tube of the length l with a sound source at its closed end and an opening at its opposite end (e.g., mouth opening). (**b**) Resonance of the closed cavity, e.g., the pulmonary airways. (**c**) Resonance of the open tube. Nodes of the sound pressure p (i.e., zero overpressure) are indicated by *filled circles* while antinodes of p (i.e., maximal overpressure and minimal underpressure) are indicated by *empty circles*; compare Footnote 26

This can be compared with the function of the vocal tract in speech production, in which the upper airways above and below the sound source (i.e., glottis) act as acoustic filters for the transmission of vowel sounds; compare Footnote 17.

In particular, an underlying vibration (or a *sound source*) generates a periodic wave of p at a specific *fundamental frequency*, whereas numerous higher harmonics usually dominate in the generated waveform (comparable with a broadband noise-like signal). As shown in Fig. 4.24a, the air-filled upper airways—in terms of *resonating cavities*—are closely attached to the sound source. The depicted *open resonating cavity* of the upper airways, i.e., *open to the outside* via open mouth and *closed* at the anatomical level of the sound source (reverberant and sound-reflecting site), *amplifies sound components* from the generated broadband signal at the particular *formant frequencies*. In fact, it is comparable with the source-filter behaviour (from Footnote 17), where the final sound emitted is determined by a product of the *sound source* and the *transfer function* of the airways.

That is, only those acoustic waves fit into the resonating cavity—i.e., resonate within the cavity and, in turn, become amplified—which fulfil *boundary conditions* of this cavity (compare Footnote 161 in Sect. 2). The latter conditions imply that a sound *pressure node* (and an antinode of the sound particle velocity) occurs at the *cavity opening* while a *pressure antinode* (and velocity node) occurs at its *closed end* where the reverberant sound source resides. Figure 4.24a demonstrates the resulting phenomenon of *standing waves*[29] within the cavity of the length l. The standing waves arise only when the axial extension l matches $\lambda/4$, $3 \cdot \lambda/4$, or $5 \cdot \lambda/4$, i.e., the sound waves fulfil the aforementioned boundary conditions. The resulting formant frequencies f_F^k of the standing waves with the index k (= 1, 2, 3,…) indicating the presence of numerous *formant frequencies* amount to

$$f_F^k = \frac{v}{\lambda} = \frac{v}{4 \cdot l} \cdot (2k - 1)\ . \tag{4.9}$$

Thus, the *transmission efficiency* of the resonating cavity in Fig. 4.24a reaches its maxima at f_F^k. To give a quantitative example, an assumed (realistic) length l of about 17 cm would yield f_F^1 of about 500 Hz and f_F^2 of about 1,500 Hz. Likewise, cavity resonances and the level of f_F^k are influenced by the shape and size of the upper airways. For the sake of completeness, it should be pointed out that the corresponding waveform of the *sound particle velocity* in the resonating cavity exhibits a phase shift of $\lambda/4$ (or 90°) in relation to the waveform of p (Fig. 4.24); compare Footnote 22.

In terms of *snoring sounds*, the lowest formant frequency (= f_F^1 in (4.9)) is related to the degree of *constriction in the pharynx*, the next higher (= f_F^2) is related to the position and shape of the *tongue*, and the one after that (= f_F^3) is correlated

---

[29] In fact, the *standing wave* within the *resonating cavity* is the sum of incident and reflected p waves which move in opposite directions (compare Footnote 170 in Sect. 2 and Sect. 6). However, the resulting standing wave oscillates only but does not propagate any more. For instance, at the closed end (*hard sound-reflecting surface*) the *incident pressure wave* $p_I = P_I \cdot \cos(kx - \omega t)$—propagating in the x direction with the pressure amplitude P_I, angular frequency ω (= $2\pi \cdot f$), and wavenumber k (= $2\pi/\lambda$)—is reflected *without phase change*. The *reflected pressure wave* $p_R = P_R \cdot \cos(kx + \omega t)$ with the amplitude $P_R = P_I = P$ interferes with p_I; e.g., interferes constructively at the closed end; compare Footnote 161 in Sect. 2. The resulting standing wave $p_I + p_R = 2P \cdot \cos(\omega t) \cdot \cos(kx)$ extends along x—with the closed end located at $x = 0$—and pulsates with t. Along the cavity in the x direction,

- *constructive interference*, i.e., amplitudes of the in-phase incident and reflected pressure waves *add*, and
- *destructive interference*, i.e., amplitudes of the out-of-phase incident and reflected pressure waves *subtract*,

occur. From a *physical point of view*, the pressure of air molecules reflecting off the closed end adds to that of air molecules approaching the closed end. In consequence, the total p doubles at the closed end, i.e., $p_I + p_R = 2P$ at $x = 0$ and $t = 0$.

with the degree of *lip-rounding* (Ng et al. 2008). For instance, the level of f_{F}^{1} was shown to be higher in obstructive apneic snores (Sect. 4.1.1.3) than in benign snores, which illustrates increasing f_{F}^{1} with increasing degree of the pharynx constriction (Ng et al. 2008), i.e., f_{F}^{1} increases with effectively decreasing l in (4.9).

As already mentioned, while sounds in the upper airways are intensified at *formant frequencies*, other sound components are damped in other specific frequency bands. For instance, *closed resonating cavities*, i.e., closed to the outside, such as closed pulmonary airways behind the sound source (Perez-Padilla et al. 1993) *absorb sounds* of specific frequencies and thus *reduce auscultatory sounds* emanating from the body. In an approximation, those sound components are damped, whose multiple half-wavelengths $k \cdot \lambda/2$ match the axial extension l of the closed resonating cavity, i.e., *pressure antinodes* should occur at both ends of the cavity, as illustrated in Fig. 4.24b. In analogy to (4.9), the resulting resonance frequencies f^{k}—also known as *harmonic eigenfrequencies*—at which sounds are absorbed can be given as

$$f^{\mathrm{k}} = \frac{v}{\lambda} = \frac{v}{2 \cdot l} \cdot k, \tag{4.10}$$

where the index k (= 1, 2, 3,...) indicates the presence of *multiple frequencies.*

For the sake of completeness, Fig. 4.24c depicts an *open tube* as a *resonating chamber*, in comparison with the *open resonating cavity* (Fig. 4.24a) and the *closed resonating cavity* (Fig. 4.24b). In the course of the *resonance* in the open tube, the sound transmission throughout the tube is most efficient. In the resonance, *pressure nodes* occur at both ends of the tube and the resulting harmonic eigenfrequencies can be calculated according to (4.10).

4.1.2.2 Effects on Sounds

After diverse propagation phenomena of body sounds have been discussed in Sect. 4.1.2.1, a highly instructive *interaction of sounds with biological tissue* should be discussed. In general, body sounds are subjected to

- *volume effects* such as *absorption*, which *attenuate* sound waves propagating in a *homogenous medium*; and
- *inhomogeneity effects* such as *scattering*, *diffraction*, *reflection*, *refraction*, and *resonance*, which *attenuate and redirect* sound waves heading in a particular direction. The latter effects are primarily caused by a *heterogeneous medium* in the sound propagation path (Fig. 4.20a).

Generally speaking, the above *effects are not fully independent* from each other. For instance, if a finite volume of tissue is exposed to ambient sounds, a part of the ambient *incident sounds* is already *reflected* back at the tissue boundary, another part of sounds is *absorbed* by tissue, and the rest is *transmitted* through this volume of tissue. In other words, the sum of reflected, absorbed, and transmitted portions of sounds should equal the ambient incident sounds. In addition, body sounds interact with biological tissues in a rather complex way so that sounds are altered not only in their intensity but also tonal quality as they pass through tissues.

Volume Effects

The aforementioned *medium-related damping* (see section "Specific Issues" in Sect. 4.1.2.1) accounts for the different *volume effects* in *homogenous medium.* That is, the absorption of sounds quantifies the *loss of sound energy* in a certain spatial direction as body sounds pass through biological tissue; consequently, *tissue is heated* up.

The process of *sound absorption* is represented by the coefficient α ((4.7) and (4.8)) considering all three (Meyer and Neumann 1975; Wodicka et al. 1989; Trendelenburg 1961; Erikson et al. 1974):

- inner friction,
- thermal conduction, and
- molecular relaxation.

Propagating sound waves are tightly interrelated with propagating (spatial) waves of p and waves of the sound particle velocity. Because of resulting differences in local sound particle velocities, an *inner friction*[30] occurs between particles oscillating with different velocity. The friction is proportional to the ratio μ/ρ. Therefore, propagation paths with stronger viscosity (or more inertial components) favour the friction and thus yield stronger damping of the propagating sound wave. The corresponding *friction-related contribution* α_F to α can be calculated as

$$\alpha_F = \frac{8\pi^2 \cdot \mu}{3 \cdot \rho \cdot v^3} \cdot f^2 , \tag{4.11}$$

whereas α_F *increases* disproportionately with the *sound frequency f.* The level of α_F in water is extremely low and amounts to about 10^{-8} m^{-1} at 1 kHz; compare Table 4.1. It can be assumed that the latter value approximately applies also to biological tissue which mainly consists of water. In comparison with water, α_F in the air is higher by three orders of magnitude[31] (Table 4.1).

In analogy with the inner friction, the propagating sound wave is linked with differences in local medium temperature. The balancing of these differences due to (finite) thermal conductivity withdraws energy from the sound wave. Likewise, the

---

[30] *Homogenous materials* tend to absorb the acoustic energy mainly because of the *inner friction*, i.e., because of local deformations and frictions within the propagation medium. In contrast, *porous materials* such as the lung parenchyma also absorb the acoustic energy in terms of the *outer friction* (Veit 1996), i.e., the friction between oscillating air particles in alveoli and semi-solid medium encircling alveoli.

[31] To give an example, if only very low values of α_F are considered (Table 4.1), the *sound pressure p* at 1 kHz would decrease by about 1 dB either after 11,000 km while *sound travelling in water*, or after 11 km while *travelling in the air* (compare the exponential term in (4.7)).

thermal conduction can be interpreted as a spatial diffusion of kinetic energy, yielding the corresponding *temperature-related contribution* α_T to α according to

$$\alpha_T = \left(\frac{c_P}{c_V} - 1\right) \cdot \frac{2\pi^2 \cdot \upsilon}{c_P \cdot \rho \cdot v^3} \cdot f^2 . \tag{4.12}$$

Here υ is the heat conductivity, whereas c_P and c_V are the specific heat capacities for constant pressure and constant volume, respectively. The level of α_T in water is three orders of magnitude lower than α_F in water. In the air, α_T is comparable in size to α_F.

The *molecular relaxation* is related to the fact that rapidly submitted sound energy—in terms of increasing local p in the propagation medium—is stored as *translational energy* of molecules, i.e., stored as increasing translational motion of molecules (solely responsible for the pressure itself). The translational energy is rapidly transferred into *rotational energy* (i.e., rotational motions of molecules) with almost no delay or, likewise, with zero *relaxation time constant* τ (≈ 0). In parallel, a part of the translational energy is converted into the *vibrational energy* (i.e., vibrational motions of molecules) *with time delay* or excitation time τ ($\gg 0$). When the local p starts to decrease within the time frame of τ (i.e., $\tau \approx 1/(2\pi \cdot f)$ with f as the sound frequency), the vibrational energy is not timely converted back into the translational energy, resulting in an *apparent loss of the instantaneous translational energy* and thus an apparent attenuation of the instantaneous local p. Likewise, the translational energy and thus the level of p appear to be greater during *compression* than during subsequent *depression*. Therefore, a part of the *sound energy seems to be lost* due to delayed relaxation; or, in other words, some of the ordered energy of the sound is transformed into random motion of the medium particles. In analogy, for relatively high sound frequencies $f \gg 1/(2\pi \cdot \tau)$, the vibrational motion of molecules is not excited (i.e., molecules can not respond fast enough). For relatively low frequencies $f \ll 1/(2\pi \cdot \tau)$, a vibrational (thermal) *equilibrium* is reached at any time, i.e., the vibrational relaxation follows in step with the sound wave. In both cases of relatively high and low frequencies, there is *no sound attenuation* by the vibrational relaxation.

The corresponding *relaxation-related contribution* α_M to α amounts to

$$\alpha_M = \left(1 - \frac{v_0^2}{v_\infty^2}\right) \cdot \frac{2\pi^2 \cdot \tau}{\left(1 + (2\pi \cdot f \cdot \tau)^2\right) \cdot v} \cdot f^2 , \tag{4.13}$$

whereas v_0 and v_∞ ($> v_0$[32]) are the sound velocities before the vibrational relaxation (i.e., for relatively low frequencies $f \ll 1/(2\pi \cdot \tau)$) and after the vibrational relaxation (for $f \gg 1/(2\pi \cdot \tau)$), respectively. It follows from the above discussion and (4.13)

[32] The *compressibility* of the propagation medium is higher at lower frequencies before the vibrational relaxation (i.e., $f \ll 1/(2\pi \cdot \tau)$) in comparison with higher frequencies after the relaxation ($f \gg 1/(2\pi \cdot \tau)$). Thus the *relation* $v_0 < v_\infty$ applies; compare the influence of D on the size of v in (4.4) (Meyer and Neumann 1975).

that sounds are subjected to a maximum loss at the *relaxation frequency* $1/(2\pi \cdot \tau)$ where the product $\alpha_M \cdot \lambda$—or the sound attenuation per sound wave cycle $v \cdot \alpha_M/f$, (4.3)—becomes a maximum.

Obviously the size of τ depends strongly on the propagation medium and tends to decrease with temperature. Complex relaxation phenomena—including chemical relaxation in terms of the ionic dissociation due to local pressure variations created by the acoustic wave—arise in liquids such as sea *water* with various dissolved substances involved. The relaxation frequency of fresh water (not sea water) is very high amounting to about $1/(2\pi \cdot \tau) \approx 80$ GHz. This high value, in turn, yields a relatively low α_M (in the range of $3 \cdot 10^{-8}$ m^{-1} at 1 kHz and 20 °C) and its strong frequency dependence ($\alpha_M \propto f^2$) considering the absorption of body sounds with their (low) frequency components only up to 2 kHz ($\ll$ 80 GHz).[33]

The relaxation phenomena are responsible for most of the acoustic losses in the *air*, whereas relaxations of oxygen and nitrogen molecules are involved. The time constant τ tends to decrease with increasing humidity because collisions of (diatomic) air molecules with water molecules favour fast transitions between different energy states; e.g., at the air humidity 70 % and temperature 20 °C the relaxation frequency is about 70 kHz for oxygen[34] and about 700 Hz for nitrogen (Rossing 2007). Likewise, the sound frequency of 1 kHz is above that associated with the relaxation of molecular nitrogen and below that associated with the oxygen relaxation. Typically, the level of α_M in the air increases with increasing water content (air humidity) and increasing sound frequency, e.g., at the humidity 70 %, temperature 20 °C, and frequency 1 kHz, the sound attenuation is about 5 dB/km or $\alpha_M = 6 \cdot 10^{-4}$ m^{-1} (> α_T, α_F of the air); compare Footnotes 31, 33 and Table 4.1.

The *total absorption coefficient* α—as used in (4.7) and (4.8)—can be given as

$$\alpha = \alpha_F + \alpha_T + \alpha_M\,. \tag{4.14}$$

It is important to observe from (4.11) to (4.13) that the level of α increases with increasing f. In particular, the contributions α_F and α_T are even proportional to f^2 while α_M is proportional to f^2 (only) below the relaxation frequency (i.e., for $f \ll 1/(2\pi \cdot \tau)$ and constant v).[35] That is, not only the propagation pathway of body sounds in the thorax

---

[33] In contrast, *sea water* shows a significantly higher α_M because of two additional relaxation phenomena in it with one relaxation frequency above 1 kHz (ionic dissociation of boric acid H_3BO_3) and another one above 100 kHz (ionic dissociation of magnesium sulphate $MgSO_4$). For instance, at the sound frequency 1 kHz and temperature 20 °C the sound attenuation in sea water totals about 0.06 dB/km or $\alpha_M = 7 \cdot 10^{-6}$ m^{-1}.

[34] For instance, the *relaxation frequency* of *pure oxygen* is only about 10 Hz yielding a large τ of about 16 ms.

[35] *Experimental data* confirm the frequency dependence of the *medium-related damping*. For instance, authors in Erikson et al. (1974) report that α is approximately proportional to f, whereas individual tissues may yield a stronger frequency dependence up to f^2, e.g., hemoglobin has α proportional to $f^{1.3}$. Studies in Loudon and Murphy (1984), Hadjileontiadis and Panas (1997a) show that the intensity of vesicular lung sounds (Sect. 4.1.1.2) declines exponentially with increasing f, which implies the proportionality between α and f; compare (4.7) and Fig. 4.22.

depends on f (Fig. 4.23) but also the medium-related damping within biological tissues increases with f, which have important practical consequences.

As illustrated in Figs. 4.1 and 4.23, the *frequency dependence of* α causes that *high frequency sounds* do not spread as diffusely or, retain as much amplitude, as do *low frequency sounds* (across the thorax). Consequently, high frequency sounds are more localised around their source. Concerning the different body sounds, it can be concluded that

- *heart sounds*—as body sounds in the low frequency range up to 100 Hz (Sect. 4.1.1.1)—are subjected to lowest attenuation in tissue, which favours their auscultation almost everywhere on the thoracic skin (Sect. 4.2.2.3).
- *Lung sounds* and *snoring sounds* experience larger attenuations than heart sounds because breathing sounds contain more high frequency components. Among them, vesicular lung sounds tend to face lowest attenuation in tissue (frequency range up to 500 Hz, Sect. 4.1.1.2) while obstructive snoring sounds face highest attenuation (range up to 2,000 Hz, Sect. 4.1.1.3). Obviously, the resulting intensity of breathing sounds on the skin (i.e., at an auscultation site) depends not only on
  - their *attenuation* in tissue but also on
  - the *frequency dependence* of their propagation pathways towards the skin,
  - the propagation *distance* to the skin, and
  - the *intensities* of their sound sources.
- *Pathological sounds*, e.g., discontinuous lung sounds, exhibit mostly high frequency components—due to a transient occurrence of such sounds—and thus do not spread as widely as *normal sounds*.[36]

From an engineering point of view,[37] the frequency dependence of α means that the transmission efficiency of the lung parenchyma and chest wall deteriorates with

[36] In fact, *high frequency sounds* exhibit *localising properties*, which are very useful in diagnosis. High frequency sounds do not spread as widely or with the intensity that low frequency sounds spread across the thorax (Ertel et al. 1966b). It means that as soon as high frequency sounds (usually pathological sounds) are heard, the corresponding sound source (or the site of pathology) is already close to the current auscultation site. This offers physicians an ability to localise pathological breathing sounds to their point of origin.

[37] *Sound transmission* through the thorax may be of high clinical value if altered transmission patterns correlate with *pathology* (Peng et al. 2014). For instance, changes in the lung structure due to the presence of *pneumothorax*—creating more barriers to the propagating acoustic waves—causes a drop in the intensity of the transmitted mechanical waves at *high frequencies* (above 100 Hz in humans (Peng et al. 2014)), which are subjected to relatively strong attenuation in tissue (see text). In contrast, sound waves at *lower frequencies* (below 100 Hz)—subjected to relatively low attenuation in tissue—can travel a longer distance (around the internal organs in the thorax) before these waves lose their energy. Consequently, structural changes of the internal organs may result in small effects on the propagation of these low frequency sounds.

Authors in Peng et al. (2014) showed that the presence of *pneumothorax* had smaller effects on the sound transmission through the thorax at *lower frequencies*. Likewise, it seems that *high frequency* mechanical waves (as could be introduced at the anterior chest surface by an actuator)

increasing *f*; i.e., *biological tissue* acts as *low-pass filter* which transmits body sounds predominantly at relatively low *f* (Wodicka et al. 1989; Welsby and Earis 2001; Welsby et al. 2003). For instance, the sound attenuation in tissue has been shown to be negligible at 100 Hz and then to increase to about 1 dB/cm at 400 Hz and even to about 3 dB/cm at 600 Hz (Wodicka et al. 1989); the corresponding level of α would amount to 11 and 34 m^{-1} at 400 and 600 Hz, respectively (compare Table 4.1).

Table 4.1 compares $\alpha_F + \alpha_T$ for different types of *physical and biological media.* It can be observed that air, adipose tissue, and lung parenchyma are strongest absorbers if the inner friction and thermal conduction are considered only. However, it should be stressed that effective values of α (4.14) are usually larger by orders of magnitude than tabled values of $\alpha_F + \alpha_T$. The contribution α_M (due to the molecular relaxation) to α is highly significant, as discussed above with regard to water and the air. Furthermore, the real absorption mechanisms in *liquids* and *semi-solids* (such as biological tissue) are highly complex; these mechanisms are also determined by *interactions* between solvent and solute and, on the other hand, governed by local *structural relaxation*, i.e., by a periodic change in the molecular arrangement due to local pressure variations created by the (mechanical) acoustic wave.

Authors in Rappaport and Sprague (1941) suggest that if effects of the inner friction (4.11) are small, as in the case with water, air, and bone, the sound energy may be transmitted with remarkably little loss. In other media, such as (breast) *fatty tissue*, sound waves are almost immediately suppressed. The flesh of the chest also acts as a strong damping medium since *obesity* might completely mask (even) *heart sounds*, i.e., sounds composed of (even) low frequency components. Likewise, relatively low frequencies of heart sounds are subjected to weak attenuation. As demonstrated in Kaniusas (2007), an increase in the *body mass index BMI* (Footnote 202 in Sect. 3) from 24 to 38 kg/m^2, i.e., an increase in obesity and amount of adipose tissue, reduced the peak amplitude of heart sounds by about 60 %.

The *attenuation of sounds* in the *lung parenchyma* deserves a few more comments. As shown in section "General Issues" in Sect. 4.1.2.1, the propagation speed *v* in the lungs is relatively slow because of the lung's elasticity dominated by a mixture of tissue and the air (Pasterkamp et al. 1997b; Kompis et al. 2001); compare Footnote 24. The non-continuous porous structure of the lung parenchyma (Footnote 30) is of special importance regarding the frequency dependence of its α. In fact, *alveoli* of the parenchyma act as *elastic air bubbles* in water, which dynamic deformations (compression and expansion) due to oscillating *p* (of the sound) dissipate the sound energy (Meyer and Neumann 1975). As long as the *size of* λ (Table 4.1) is significantly greater than the alveolar size (diameter < 1 mm), the sound losses are relatively low.

(Footnote 37 continued)
propagate directly (to the posterior chest surface where a sensor resides) through internal organs (lying between the actuator and sensor). Therefore, any change in the intrathoracic structure would affect the propagation of *high frequencies* through the thorax.

Here the arising losses due to the thermal conduction[38] are considerably larger than those associated with the inner friction (viscous effects) and scattering effects (Wodicka et al. 1989). When the size of λ approaches the alveolar size, i.e., the sound frequency is increasing (4.3), the losses start to increase strongly (Pasterkamp et al. 1997b). However, it is important to observe from Table 4.1 that typical values of λ in the lung parenchyma—despite the relatively low v—are still significantly larger than the alveolar size if the typical frequency range of body sounds (up to 2 kHz) is considered. Provided that v = 23 m/s (as the lowest reported value from section "General Issues" in Sect. 4.1.2.1), the resulting size of λ would approach the alveolar size at the earliest at f = 23 kHz.

From an engineering point of view, *low-pass behaviour of the lung* can be expected because damping of sounds in tissue increases with frequency. That is, low frequency components of lung sounds are predominantly transmitted through the lung while high frequency components are filtered out by the lung. As reported in Fachinger (2003), the cut-off frequency of such low-pass behaviour is at about 400 Hz. Interestingly, the *cut-off frequency decreases* with an increasing *accumulation of the air* in the lungs (reduced density of the lungs), which impedes the transmission of lung sounds. On the other hand, increased density of the lungs—in terms of *parenchymal consolidation* as can be caused by *illness—increases the cut-off frequency* and thus facilitates the sound transmission (especially, high frequency sound components).

The *consolidated lung* acts as an efficient sound conductor. An increased volumetric portion of tissue in the lungs, e.g., in the congested lungs (compare Footnote 24), favours the transmission of voice sounds to the chest wall, *especially at high sound frequencies* (Wodicka et al. 1989). In fact, the consolidation may cause important changes in the quality of lung sounds recorded on the chest wall. For instance, sounds originating in bronchial and tracheal tracts would be well transmitted to the *chest wall*, a distant sensing site in relation to (centrally located) sound sources, which is a consequence of improved sound transmission through the consolidated parenchyma. In contrast to effects of the *normal lung*—yielding (by definition) *vesicular lung sounds* on the chest—auscultated *sounds on the chest* would be similar in quality to *tracheobronchial lung sounds*, i.e., would be relatively loud and high-pitched (Sect. 4.1.1.2), especially during the expiration phase in which normal vesicular sounds are absent. In addition, *expiratory sounds* would be as loud as *inspiratory sounds*, which is contrary to the quality of vesicular sounds. Likewise, tracheobronchial sounds if heard on the chest instead or in addition to vesicular sounds indicate pathologically consolidated lungs (Loudon and Murphy 1984; Dalmay et al. 1995; Wodicka et al. 1989). Finally, it should be

[38] In this case, *thermal losses* arise because *bubble compressions* require greater work performed by the acoustic wave than the work performed by the air in bubbles during *bubble expansions* (Wodicka et al. 1989). The resulting energy difference is conducted into the lung tissue as heat. Interestingly, *enlarged alveoli* tend to increase thermal losses and thus to attenuate more strongly body sounds within the lungs in comparison with *reduced alveoli* (pre-compressed bubbles).

mentioned that an experimental estimation of the transmission characteristics of body sounds in the thorax can even lead to diagnosis and categorization of (respiratory) diseases because different diseases affect the transmission characteristics in a unique way.

Inhomogeneity Effects

In addition to the geometry-related damping of propagating body sounds (see section "Specific Issues" in Sect. 4.1.2.1) and the sound absorption in terms of the medium-related damping in *homogenous media* (see section "Volume Effects" in Sect. 4.1.2.2), the *heterogeneous structure* of biological tissues impacts strongly the *attenuation of sounds* propagating in a specific direction, e.g., towards the skin surface. A highly heterogeneous structure of the thoracic region is demonstrated in Fig. 4.20a. In general, body sounds interact with such structures in a rather complex way and may experience *spatial redirection*, *accumulated attenuation* with respect to an acoustical sensing device (on the skin), and even *local amplification* because of resonating effects in spatially delimited cavities.

In particular, the following *effects govern the propagation of body sounds* in tissue towards the skin surface and the coupling of sounds into the air above the skin (or into the sensing device, Fig. 4.1):

- scattering,
- diffraction,
- reflection,
- refraction, and
- resonance.

Scattering and Diffraction

Sound waves are *scattered*, i.e., redirected in random directions from a straight trajectory, when waves encounter inhomogeneities[39] (or obstacles) in the propagation medium; see a detailed discussion on scattering in section "Inhomogeneity effects" in Sect. 5.1.2.2. Such *inhomogeneities* in the thorax are given, for instance, by spatially delimited inner organs, airways, bones, skin, and—on a smaller scale—different types of tissues, porous lung parenchyma, (large) blood vessels. The dimensions of the former *structures are nearly in the range of* λ (Table 4.1) while those of the latter structures are already much smaller than λ.[40]

---

[39] From an acoustical point of view, *inhomogeneities* or *obstacles* are given by media with different Z (4.6). That is, fluctuations of the medium *density* ρ or the varying *propagation velocity* v of sounds (when entering a different medium, (4.4)) constitute inhomogeneities for the sound wave.

[40] *Inhomogeneities* on an *even smaller scale* such as cellular structures or protein aggregates are unimportant for the scattering of body sounds because the effective λ of sounds is already orders of magnitude larger than the dimensions of these inhomogeneities. However, such small structures are highly relevant for the *optical scattering* (Sect. 5.1).

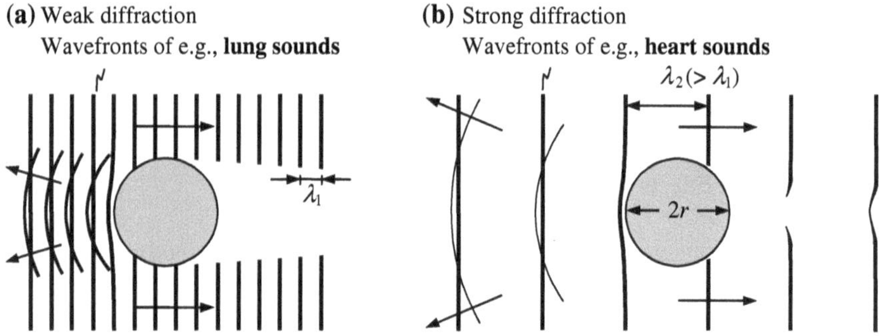

Fig. 4.25 Scattering and diffraction of body sounds off of an obstacle at (**a**) relatively high frequencies ($\lambda_1 < 2r$, see (4.3)) and (**b**) relatively low frequencies of sounds ($\lambda_2 \approx 2r$ and $\lambda_1 < \lambda_2$); compare Sect. 6. Here λ denominates the sound wavelength and $2r$ the diameter of the obstacle. The *line thickness* of the wavefront indicates the incident, scattered, and diffracted sound intensities, whereas the *arrows* indicate the propagation direction

Generally, some part of the sound wave is *scattered* (or reflected in many directions) off of an obstacle, another part is *absorbed* (see section "Volume Effects" in Sect. 4.1.2.2), whereas the rest of the wave propagates *around the obstacle*, as illustrated in Fig. 4.25. The *bending of the sound wave* around *small obstacles*—small compared to the size of λ or on the order of λ—is coined as wave *diffraction* (Fig. 4.25b).

- If the size of λ *is large* in relation to the obstacle, only a small part of the wave will be scattered, i.e., only the tiny part that strikes the obstacle; the larger part of this wave will *readily diffract* around the obstacle and remains unaffected (Fig. 4.25b). Likewise, for longer λ the obstacle behaves as a point source of diffracted waves and the resulting *scattering losses* are small, whereas the obstacle's shape is of little importance. In analogy, the sound can also spread out beyond *small openings*,[41] i.e., small compared to λ; one could consider them to be anti-obstacles, which then act as localized point sources of sound.
- In contrast to large λ, if the size of λ *is in the same order* of magnitude as the size of the obstacle, the (back) scattering tends to increase while the diffraction (bending towards the obstacle of waves past this obstacle, Fig. 4.25b) becomes less pronounced. The *interference* may arise among diffracted waves creating (alternating)

[41] In fact, every unobstructed point on the *incident wavefront* momentarily present in the *opening* (or slit) acts as a source of a *secondary spherical wave*. The superposition of all spherical waves determines the form of the *resulting transmitted wavefront* at any subsequent time behind the slit, i.e., the *superposition* determines the resulting *diffraction pattern* of the slit. Obviously, not only amplitudes but also relative phases of the individual spherical waves govern their *interference pattern* and thus the resulting transmitted wavefront beyond the opening. Namely,

- *in-phase superposition* leads to *constructive interference* and thus to the maximum of the transmitted intensity at a certain observation point beyond the opening. In contrast,
- *out-of-phase superposition* leads to *destructive interference* and thus to the null in the transmitted intensity at an observation point beyond the opening; for details see Sect. 6.

regions of greater sound intensity (known as constructive interference) and lesser sound intensity (destructive interference); compare Footnote 41.

- For an even *smaller* λ below the size of the obstacle, most of the sound incident on the obstacle is scattered (back) according to the *reflection laws* (4.15), the *diffraction almost disappears*, and a *sound shadow*[42] is formed behind this obstacle (Fig. 4.25a). Likewise, the back scattered energy is much larger for the reflection than diffraction, as illustrated schematically in Fig. 4.25a, b.

Thus obstacles smaller than λ tend to *scatter in all directions* while obstacles bigger than λ tend to *scatter more directional*; compare Fig. 4.25a with Fig. 4.25b (see also Fig. 5.11). In fact, multiple scattering-related redirections—or *multiple scattering events*, compare Fig. 5.10a—contribute to the *overall attenuation* of the propagating sounds when arriving at a distant location, e.g., at a distant sensing device on the skin.

This yields that *high frequency sounds*, with relatively short λ (4.3), do not diffract around obstacles, but are predominantly absorbed (due to their relatively high frequency, section "Volume Effects" in Sect. 4.1.2.2) or reflected instead (see below), which creates the *sound shadow* (Fig. 4.25a). In contrast, *low frequency sounds* have relatively long λ which usually exceeds the dimensions of obstacles in the propagation pathway and thus are bent around these obstacles. These sounds diffract and *pass around obstacles undisturbed* so that the sound wave—already several wavelengths past the obstacle (Fig. 4.25b)—is fully identical with that in front of the obstacle. Likewise, high frequencies scatter much more than low frequencies.

Considering particular body sounds, it can be concluded that *heart sounds*, i.e., low frequency sounds, tend to spread *diffusely* in all directions and diffract more strongly than *lung sounds* or *snoring sounds*, i.e., high frequency sounds, which spread more *directional*. From a practical point of view, it favours the auscultation of heart sounds everywhere on the chest skin.

Reflection

Provided that the dimensions of the obstacle—in the sound propagation pathway—are larger than λ of the sound wave, the wave is coherently *reflected* off of the obstacle at the *boundary* according to the *reflection laws*. Namely, the *incident angle* to the normal (to the reflective surface at the point of the incidence) equals the *reflection angle* to the normal, whereas all three the incident wave, the reflected wave, and the normal lie in the same plane. Figure 4.26 demonstrates the reflection of inner body sounds—emanating from the body—on the inner skin surface, i.e., on

[42] It is interesting to note that the *boundary* between the *sound wave* (i.e., compressions and rarefactions) and the *sound shadow* (i.e., died wave) always extends over a certain number of wavelengths because the mechanical sound wave can not die abruptly due to elastic interactions among adjacent molecules. This effectively determines the spatial extension of the diffraction, which is greater at large λ (or low sound frequency) and less at small λ (or high sound frequency). Likewise, a sound shadow behind an obstacle decreases in size with increasing λ.

the *discontinuity tissue-air*. The equality of the incident angle φ_T and the reflection angle φ'_T can be observed.

The amount of the reflected wave (as related to the incident wave) is determined by the *acoustic reflection factor* Γ_A (compare 2.32) given by

$$\Gamma_A = \frac{P_R}{P_I} = \frac{Z_A - Z_T}{Z_A + Z_T}. \tag{4.15}$$

Here P_R and P_I are the respective amplitudes (= peak values) of the reflected and incident (pulsatile) sound pressure; compare Footnote 29. The *characteristic acoustic impedance* Z_A denotes the air impedance above the skin while Z_T denotes the tissue impedance below the skin. An approximate estimation of Γ_A with (4.6), data from Table 4.1 and Footnote 24 yields $Z_A = 343\ \text{kg} \cdot \text{m}^{-2} \cdot \text{s}^{-1}$, $Z_T = 1.5 \cdot 10^6\ \text{kg} \cdot \text{m}^{-2} \cdot \text{s}^{-1}$, and a very large $\Gamma_A = -0.99$. The minus sign of Γ_A indicates that the reflected sound (pressure) wave experiences a *phase reversal* in relation to the incident sound wave. This is because the *air comprises a very soft medium* in comparison with the tissue (i.e., $Z_A \ll Z_T$); compare Footnotes 29 and 36 in Sect. 5. Consequently, the total sound pressure (= $p_I + p_R$) at the tissue-air interface has to drop to fulfil the boundary condition ($p_I + p_R \rightarrow 0$) so that the resulting reflected wave has to satisfy $p_R \approx -p_I$. Figure 4.26 also illustrates the phase reversal by space-shifted incident and reflected wavefronts in the tissue below the skin. The discontinuity tissue-air is partially comparable with the effect of resonance cavity opening in Fig. 4.24a, c; compare also Footnote 36 in Sect. 5.

On the other hand, the high value of $|\Gamma_A|$ would indicate that more than 99 % of the incident wave is reflected back on the inner skin surface while only 1 % is transmitted through the skin and is then available for the acoustical sensing device. Likewise, the acoustic *impedance mismatch* between different body tissues always accounts for the *sound attenuation* while the associated sound wave crosses boundaries. On the other hand, matched impedances will not yield any attenuation ($\Gamma_A = 0$ for $Z_A = Z_T$ in (4.15)). For instance, the impedance mismatch between the lung parenchyma and chest wall can account for a significant decrease in p by an order of magnitude because the chest wall is much more massive and stiffer than the parenchyma (Pasterkamp et al. 1997b).

However, the above estimation of such high *reflection losses* is of limited validity because the *reflection laws* (including (4.15)) hold only when λ of the sound is relatively small. As already discussed, the latter requirement is hardly met by body sounds in the frequency range up to 2 kHz (Table 4.1), the effective λ being larger or at most equal to the dimensions of inner body structures. In other words, the estimated high reflection losses are rather valid for *high frequency body sounds* with correspondingly short λ. Figure 4.26 signifies the loss in the intensity of body sounds while crossing the simplified tissue-air boundary.

In addition, the above estimation assumes a *simplified tissue-air interface* with a single boundary, whereas the *real skin* constitutes a true *multilayer* consisting of at least three layers, namely, the innermost subcutaneous fat, the dermis, and the outermost epidermis. Thus, there are at least four boundaries for body sounds to

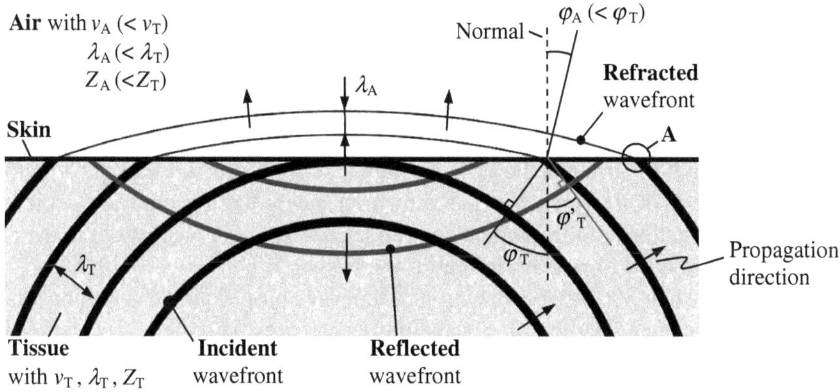

Fig. 4.26 Reflection and refraction of body sounds on the tissue-air boundary with v_T and v_A as the sound propagation velocity in the tissue and air, respectively. Here λ is the corresponding sound wavelength, Z the corresponding characteristic acoustic impedance, φ_T the incident angle in the tissue, φ'_T (= φ_T) the reflection angle in the tissue, and φ_A the refraction angle in the air. The *line thickness* of wavefronts indicates roughly the sound intensity

cross while transversing the skin. It can be assumed that *neighbouring layers*—exhibiting similarities in their physical and thus acoustical properties—show *less difference in their Z* than the difference $|Z_A - Z_T|$ of the simplified tissue-air interface. Therefore, the effective Γ of boundaries in the multilayer would be less in comparison with the estimated Γ_A of the simplified two-layer system.[43]

Refraction

The part of sounds, which was not reflected at the boundary in the propagation pathway, enters another medium behind the boundary (Footnote 39) and usually experiences a redirection (or *bending*) owing to a *change in v*, known as wave *refraction*. In contrast to scattering, diffraction, and reflection, involving spreading and bending of sound waves in a *single medium* with (almost) constant *v*, refraction involves bending of sound waves which enter *another medium* with different *v*.

[43] For instance, if a *trilayer* is assumed with only one intermediate layer (with Z_I) between the tissue (Z_T) and the air (Z_A), the resulting two *reflection factors* Γ_M (of two reflecting surfaces) for sounds emanating from the body would amount to

$$\Gamma_{M,1} = \frac{Z_I - Z_T}{Z_I + Z_T} \quad \text{and} \quad \Gamma_{M,2} = \frac{Z_A - Z_I}{Z_A + Z_I}.$$

Provided that $Z_A < Z_I < Z_T$, the respective magnitudes of Γ_M satisfy $|\Gamma_{M,1}| < |\Gamma_A|$ and $|\Gamma_{M,2}| < |\Gamma_A|$, whereas Γ_A of a simplified *bilayer* tissue-air is given by (4.15). Thus the trilayer shows lower reflection losses in comparison with the *bilayer*.

As illustrated in Fig. 4.26, body sounds in the tissue (*fast medium* with v_T) approach the air where their speed v_A is slower (Table 4.1). As soon as the incident wavefront hits the *slow medium* the wavefront is slowed down at the point of the incidence, for instance, in the region A in Fig. 4.26. Meanwhile, the rest of the wavefront in the tissue, e.g., to the right of the region A, continues to spread with a high speed v_T until it also hits the slow medium. Consequently, the *wavefront* in the region A is *bend* to the left when the wavefront enters the air; compare depicted propagation directions of the incident and refracted waves. The wavefront in the air seems to be *flattened* in relation to that in the tissue.

It is obvious that not only the direction of the wave propagation changes but also the distance between neighbouring wavefronts decreases, i.e., the *wavelength* λ is decreased in the air related to the incident λ in the tissue (Fig. 4.26). Likewise, a lower v yields a shorter λ in a medium (4.3) because the sound *frequency* f does not change from one medium to another (in linear media only); the level of f is solely determined by the rate of mechanical vibrations in the sound source.

As in the case of reflection, the laws of *acoustic refraction* apply only if the dimensions of the refracting surface are larger than λ of the sound. Namely, the *refraction angle* φ_A (in the air) and the incident angle φ_T (in the tissue) obey *Snell's refraction law*[44]:

$$\frac{v_A}{v_T} = \frac{\sin(\varphi_A)}{\sin(\varphi_T)}, \tag{4.16}$$

whereas the incident wave, the refracted wave, and the normal lie in the same plane (Fig. 4.26). Since the inequality $v_A < v_T$ applies (Table 4.1), the angle φ_A is smaller than φ_T. Likewise, the refracted wave of inner body sounds is bent towards the normal of the skin, i.e., the air wavefront becomes flattened.

It should be noted that the *flattened wavefront* in the air *favours the auscultation* of body sounds via a (usually) flat acoustical sensing device on the skin (Fig. 4.1). In particular, this advantage is rather applicable for *high frequency body sounds* with correspondingly short λ.

Resonance

Some *inhomogeneities* in the thorax, especially in the respiratory tract, build *tube-like resonating cavities* such as the *air-filled* upper airways and pulmonary airways. As described in section "Specific Issues" in Sect. 4.1.2.1, cavities act as resonating

---

[44] Willebrord Snellius (1580–1626) was a Dutch astronomer and mathematician after which *Snell's law* was named. This law relates the degree of the wave bending to the physical properties of materials which surround the bending surface.

acoustic filters. Here *incident waves* reflect at boundaries of a cavity and then interfere with *reflected waves* formed in this way. It leads to *constructive and destructive interference* along the resonating cavity and, in turn, to the phenomenon of *standing waves* in the cavity; compare Fig. 4.24a, b and Footnote 29. It should be noted that such resonating effects are less likely in semi-solid tissues of the body, even though tissues are spatially delimited by e.g., skin or airway walls. The reason for this is that the level of λ in tissues (Table 4.1) is larger than the corresponding dimensions of (homogenous) tissues.

4.2 Sensing Aspects

According to Figs. 4.1 and 4.2, *sensing aspects* include

- *coupling* of body sounds from the skin into an acoustical sensing device applied on the skin and
- *conversion* of the sound pressure of coupled sounds into an electric signal within the sensing device.

In particular, *body sounds*—mechanical vibrations (and forces) arrived on the skin surface from the inner body—induce mechanical *vibrations of the skin surface*.[45] These vibrations serve as sound sources accessible to the *sensing device* (Fig. 4.1). Profound understanding of diverse coupling aspects offers a solid basis for the *interpretation and diagnostic use of auscultated body sounds*, facilitating also the optimisation of sound auscultation techniques.

As illustrated in Fig. 4.1, a *mixture of body sounds* arrives at the skin level, composed of mainly *heart sounds*, *lung sounds*, and *snoring sounds* (Sects. 4.1.1.1–4.1.1.3). Since different and (almost) independent sound generation mechanisms underlay the different body sounds, it can be assumed that auscultated sounds on the skin represent a mixture of *additive contributions* from the different body sounds. In addition, there are numerous *mutual interrelations* between the different body sounds, as already discussed in Sect. 4.1.1.5.

[45] In fact, body sounds cause *skin vibrations* of three different waveform types: *transverse waves* (or shear waves), *longitudinal waves* (or compression waves, compare Fig. 4.19), and a combination of the two types (Ertel et al. 1971). The corresponding *deflection amplitude* of particles involved in the transmission of *acoustic sounds*, e.g., the deflection amplitude of air molecules while transmitting air sounds (Footnote 22), is proportional to the *sound pressure* level and inversely proportional to the *sound frequency*, medium density, and sound velocity (Giancoli 2006). To give a quantitative example, the *deflection in air* at 1 kHz is about 8 μm at the sound *threshold of pain* in humans and less than 0.1 nm (i.e., the approximate size of an atom) at the *threshold of human hearing*.

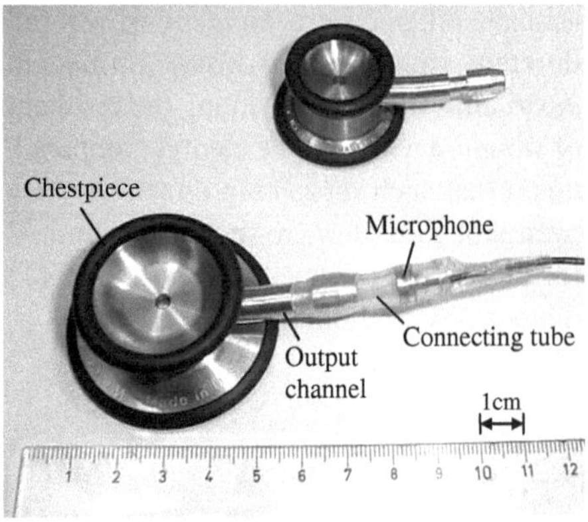

Fig. 4.27 Body sound sensors. Small and large stethoscope chestpieces (Fig. 4.31) for the auscultation of high and low frequency sounds, respectively; compare Fig. 4.29

4.2.1 Coupling of Body Sounds

It is the *chestpiece* of a standard stethoscope—as illustrated in Fig. 4.27—which forms the basis of the acoustical sensing device[46] (Kaniusas et al. 2005). The chestpiece collects the different *body sounds* by converting mechanical vibrations of body *tissues* of the chest wall (or vibrations of the *skin surface*) into sound vibrations of the *air* within the chestpiece (Fig. 4.1). The airy output of the chestpiece is targeted at a microphone in order to establish an electrical *output signal* proportional to vibrations of the sound pressure at the output.

Body sounds are strongly altered in their properties by *imperfect coupling* from the tissue, throughout the skin, and into the air of the chestpiece. Physical properties of this *acoustical transmission path* have strong implications on the filtering of body sounds; some sound components are attenuated while others are amplified. Dominant *viscoelastic properties* (Footnote 134 in Sect. 2) of the skin render the interaction between tissue-bound body sounds and the resulting skin vibrations even more complex.

It will be shown that the *chestpiece* (Sect. 4.2.1.1) can be considered as an acoustic amplifier with a specific resonance behaviour, whereas the *microphone* (Sect. 4.2.1.2) serves as an (almost ideal) electro-acoustic converter. In addition, the chestpiece is an integrative part of the widely used *stethoscope* (Sect. 4.2.1.3) and thus determines strongly its acoustical transmission.

[46] For the sake of completeness, it should be noted that there are other *acoustical sensing devices*, besides the chestpiece. For instance, piezoelectric sensors shaped as a flat diaphragm can also be used for direct skin attachment and the recording of body sounds.

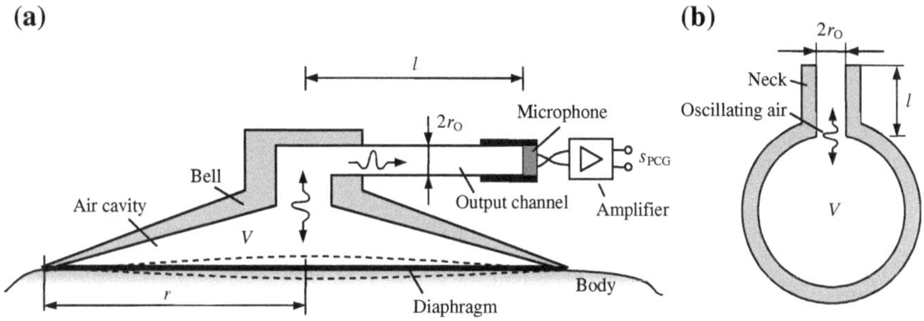

Fig. 4.28 (**a**) The structure of the stethoscope chestpiece applied on the skin for the auscultation of body sounds, namely, for the establishment of an acoustic biosignal phonocardiogram s_{PCG}; compare Fig. 4.27. (**b**) The cross section of Helmholtz resonator which resembles the bell of the chestpiece

4.2.1.1 Chestpiece

The *chestpiece* consists of a *circular diaphragm* and a *shallow bell*, as illustrated in Fig. 4.28a. The *acoustic* transmission path of the chestpiece begins with the diaphragm which is in close contact with the skin. The diaphragm follows the *vibration of the skin*[47] which, in turn, follows mechanical forces of inner body sounds. This vibration of the diaphragm creates mechanical *pressure waves* within the air of the bell. The waves travel across the bell towards the output of the chestpiece where a microphone is attached. Consequently, an oscillation of the *microphone membrane* is induced, which finally yields an *electric output signal* (Sect. 4.2.1.2). The diaphragm may also be absent in the chestpiece, as will be discussed later.

Diaphragm

The vibration of the circular *diaphragm* (or membrane) exhibits many *vibrational modes* in which all space elements of the membrane oscillate with the same frequency and fixed phase relation. The *fundamental oscillation mode* of the diaphragm is demonstrated in Fig. 4.28a, where the diaphragm's midpoint shows the highest deflection amplitude and the entire diaphragm moves in-phase. In fact, the maximum deflection amplitude (or the maximum sensitivity of the diaphragm) in

[47] The *vibration amplitude of the skin*—may be less than a few μm, compare Footnote 45—depends strongly on the method of sound recording. For instance, a *massive chestpiece* and a *tight skin contact* would impose a significant mechanical loading on the skin surface. Consequently, the resulting mechanical stress would rise in the skin beneath the chestpiece, which would significantly *limit the mechanical deflection amplitude* of the skin surface; compare with the influence of the *pre-stressed skin* on the acoustic transfer function of the chestpiece (Fig. 4.29 and Footnote 52).

response to an *external excitation* (e.g., to emanating body sounds) occurs at the mechanical *resonance of the diaphragm*, i.e., at the mechanical resonance frequency of the diaphragm. Each mode has its own resonance frequency, known as *eigenfrequency* f_{km} with k and m as mode numbers, according to

$$f_{\mathrm{km}} = \frac{c_{\mathrm{km}}}{2\pi \cdot r} \cdot \sqrt{\frac{\sigma}{\rho}} = \frac{c_{\mathrm{km}}}{2\pi \cdot r} \cdot v \,. \tag{4.17}$$

Here c_{km} denotes coefficients for the vibrational mode (k, m), whereas c_{km} is given by the mth zero of Bessel[48] function of the order k. Furthermore, r is the diaphragm radius, ρ the density of the material the diaphragm is made of, v the sound propagation velocity along the diaphragm of transverse sound waves, and σ the *mechanical stress* in the diaphragm (due to non-zero surface tension).

It is important to note that f_{km} represents a series of *non-harmonic eigenfrequencies* while the size of f_{km} increases with increasing k and m. The fundamental oscillation mode (0, 1) has the lowest $f_{\mathrm{km}} = f_{01}$ with $c_{\mathrm{km}} = c_{01} = 2.4$. It can be expected that *only the fundamental mode* (0, 1) is excited in the diaphragm of the chestpiece (Fig. 4.28a), as also indicated in Rappaport and Sprague (1941). This is because of a relatively tight skin contact and, on the other hand, relatively high values of f_{km} of higher modes, potentially exceeding the frequency range of body sounds (up to 2 kHz). Thus only f_{01} is practically relevant.

It can be observed from (4.17) that *eigenfrequencies* f_{km} *increase* with

- *decreasing* r, i.e., with the diaphragm getting smaller, and, on the other hand, with
- *increasing* σ, i.e., with the rising surface tension (or the pre-stress) of the diaphragm.

Both effects have important *diagnostic consequences*, as will be discussed later.

In fact, the *skin* encompassed by the rim of the bell acts as a *natural diaphragm*, even if the diaphragm of the chestpiece is removed (Fig. 4.28a). The encompassed skin behaves with the properties of a *damped diaphragm* of relatively low v (from (4.17)). Provided that the *artificial diaphragm* covers the chestpiece, the circular skin region is in tight mechanical contact with this diaphragm and *oscillates concurrently* with the latter. As a result, the *effective diaphragm* consists of

- the artificial diaphragm and
- the natural diaphragm,

[48] Friedrich Wilhelm Bessel (1784–1846) was a German astronomer who systematically derived Bessel functions appearing in mathematical descriptions of many physical phenomena, such as the flow of heat or the propagation of electromagnetic waves.

both governing significantly acoustical properties of the chestpiece, especially, of the skin-diaphragm interface (Rappaport and Sprague 1941; Hollins 1971); compare Footnote 52.

Bell

The diaphragm of the chestpiece is clamped by the *bell* which is usually shaped as a funnel and operates as a *resonating cavity*. In contrast to the resonating cavities from Fig. 4.24, the dimensions of the bell are smaller than the size of λ (in the air, Table 4.1) so that *no standing waves* can be formed in the bell; to be more precise, no dimension of the resonator exceeds $\lambda/4$. In fact, the bell serves as a *transducer* which converts the vibration of the diaphragm (i.e., the effective diaphragm or the natural only) into longitudinal sound waves propagating through the air within the cavity (Ertel et al. 1971). Likewise, the resonating cavity is *excited by the diaphragm*, *absorbs* best a band of sound frequencies around its resonating frequency, and transmits the absorbed sounds towards its output channel (Fig. 4.28a).

As illustrated in Fig. 4.28, the bell with its output channel resembles *Helmholtz*[49] *resonator*, a container of air (or cavity) with a short small-diameter open neck.[50] As long as the dimensions of Helmholtz resonator are smaller than $\lambda/4$, i.e., there is no significant pressure variation inside the cavity, a single *resonance frequency* f_{HR} dominates:

$$f_{HR} = \frac{v}{2\pi} \cdot \sqrt{\frac{A}{V \cdot l'}}\,. \tag{4.18}$$

Here v is the sound propagation velocity in the air, A the cross sectional area of the opening port (or the output channel), i.e., $A = \pi \cdot r_O^2$ with r_O as the radius of the opening port, V the inner volume of the resonator, and l' the effective length the oscillating air column has in the opening port. The size of l' depends on r_O and is usually approximated as $l' = l + \pi/2 \cdot r_O$ (Veit 1996) with l as the geometrical length of the opening port (Fig. 4.28).

---

[49] Hermann von Helmholtz (1821–1894) was a German scientist and philosopher whose groundbreaking investigations occupied almost the whole field of science, including physiology, physics, electricity, and chemistry.

[50] The function of *Helmholtz resonator* can be summarised as follows. A volume of the air in and near the neck—compare Fig. 4.28b—starts to vibrate in response to *external excitation*. For instance, pushing extra air down the neck into the cavity creates an overpressure in the cavity. After release of the external force, the air rushes out due to the *springiness* (or compressibility) of the air within the cavity. Shortly afterwards, the air pressure inside the cavity undershoots the equilibrium level (i.e., the atmospheric pressure, Footnote 26) because the air in the neck has mass and thus possesses *momentum* when it rushes out. A slight vacuum occurs in the cavity, which then sucks some air back into the cavity. It results in a (damped) oscillation of the air (in and near the neck) into and out of the cavity at a specific *natural frequency*, known as the *resonance frequency* f_{HR}.

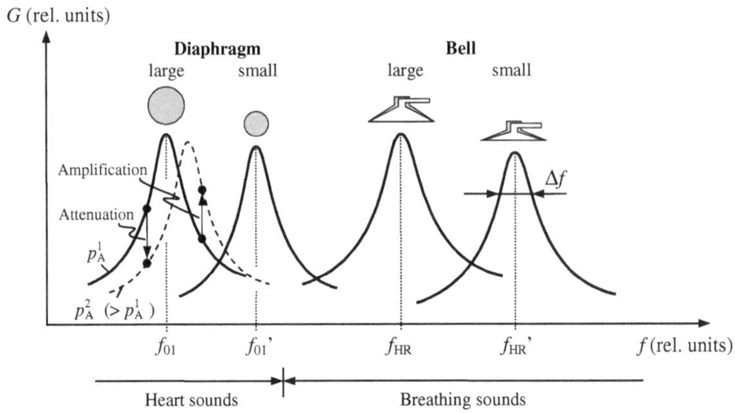

Fig. 4.29 Contributions of the chestpiece bell (Helmholtz resonator) and the chestpiece diaphragm (Fig. 4.28a) to the acoustic transfer function G of the chestpiece; compare Fig. 4.32. Large and small bells are compared with each other as well as large and small diaphragms (considering only the fundamental oscillation mode with the eigenfrequency f_{01}, see (4.17)). The impact of increased application pressure p_A of the chestpiece (with the diaphragm) on the skin is also indicated for $p_A^1 < p_A^2$. Approximate frequency regions of heart sounds and breathing sounds are depicted for comparison

It can be observed from (4.18) that the *frequency* f_{HR} *increases* with

- *increasing* A because a larger opening accelerates the escape (and the inflow) of the air out of (and into) the cavity, leading to a higher f_{HR};
- *decreasing* V because less air must move out (and into) the cavity to relieve (and build up) the sound pressure within the cavity towards (and away from) the atmospheric pressure (Footnote 26), leading to faster response times; and
- *decreasing* l (or l') because a shorter opening port induces less resistance to the oscillating air flow and decreases momentum (inertia) of the oscillating air in the port, leading to a higher f_{HR}.

Figure 4.29 depicts schematically the *acoustic transfer functions* (or frequency response curves) of the chestpiece *diaphragm and bell*. It is obvious that the diaphragm transmits optimally body sounds into the bell at its f_{01} (4.17). Sound components below and above f_{01} are damped because the diaphragm is excited at these frequencies to a lesser extent. Since the *vibration of the diaphragm* is damped, the corresponding resonance curve is widened around f_{01}, whereas the widening is quantitatively described by the *quality factor*.[51] It can be observed in Fig. 4.29 that

[51] The *quality factor* represents the degree to which an *oscillatory system* is undamped. It is defined as the ratio of the *resonance frequency* (e.g., f_{HR} from (4.18)) to the *bandwidth* Δf of the sound oscillation. Here the bandwidth Δf is determined as the difference of two frequencies (above f_{HR} and below f_{HR}) at which the acoustical power left (or dissipated) in the oscillatory system is one-half (or 3 dB less than) its maximum value at f_{HR}, compare Fig. 4.29. The quality factor > 0.5 represents an *underdamped system* in which oscillations can arise in response to an *external*

the resonance curve moves towards higher sound frequencies with decreasing r (or increasing σ). This means that smaller diaphragm or an increased *application pressure*[52] *of the chestpiece* on the skin facilitates attenuation of lower frequencies ($< f_{01}$, Fig. 4.29) and amplification of higher frequencies ($> f_{01}$). In analogy, increased mounting surface tension of the diaphragm (clamped by the bell) or a static excursion (*pre-bending*) of the diaphragm while applied on the skin increases σ and thus raises f_{01}, e.g., the eigenfrequency f_{01} moves toward f_{01}' (Fig. 4.29). Likewise, the auscultation of *high frequency body sounds* or, in analogy, the reduction of *low frequency body sounds*, is favoured by small r and high σ, which is highly significant in diagnosis[53] of body sounds. In addition, the reduction of low frequency body sounds tends to unmask high frequency body sounds. Usually low frequency sounds are relatively loud and are audible in the first place, *masking*[54] the informative high frequency sounds (usually faint but pathologic) when presented together.

This means that the *attenuation of low frequencies* or *unmasking of high frequencies* is favoured by *decreasing r*; compare Fig. 4.29 and (4.17). In addition, the efficiency of a deliberately reinforced unmasking by an increased application pressure improves with decreasing r. This is because the tendency of the application

---

(Footnote 51 continued)
disturbance displacing the system from its equilibrium state. Otherwise, the factor < 0.5 represents an *overdamped system* and implies an exponential decay back to the equilibrium state in response to a temporal external disturbance. In other words, lossy materials have lower quality factor and make the response curve (transfer function) wider and lower; i.e., Δf increases while G decreases in Fig. 4.29.

[52] The *application pressure of the chestpiece* on the skin is a relevant issue because the varying pressure alters sound filtering characteristics of the diaphragm (Fig. 4.28). In fact, the chestpiece must be used with the principles of a *damped diaphragm* in mind, namely,

- the *lightest* possible application for the auscultation of *low frequency sounds* and
- the *firmest* possible application for the auscultation of *high frequency sounds.*

It is important to note here that increasing application pressure increases not only the pre-stress of the artificial diaphragm but also the *mechanical stress of the skin* region under the diaphragm, the skin encompassed by the rim of the bell. As a result, increased stress of the skin—or increased pre-stress of the *natural diaphragm* so formed (4.17)—contributes to the auscultation of high frequency sound components because the natural diaphragm oscillates concurrently with the *artificial diaphragm.*

[53] From a diagnostic point of view, the effect of the *attenuation of lower frequencies* ($< f_{01}$, Fig. 4.29) by simply increasing the *application pressure of the chestpiece* on the chest wall is deliberately used by physicians (Hollins 1971; Ertel et al. 1966b; Abella et al. 1992); compare Footnote 52. That is, increased application pressure favours the detection of high frequency sounds ($> f_{01}$, Fig. 4.29) which usually indicate *pathology* (see section "Volume Effects" in Sect. 4.1.2.2) and possess *localising properties* (Footnote 36).

[54] *Masking of body sounds* is important in two respects (Rappaport and Sprague 1941),

- masking of a sound in the presence of other sounds and
- masking of a sound following another sound of considerably large intensity.

In the first case, as a sound mixture becomes more intense, the low pitched sound components (e.g., *heart sounds*) start to dominate because the high pitched components (e.g., *lung sounds*) are masked by peculiar characteristics of human hearing. In the second case, a preceding sound of a comparably great intensity tends to temporarily fatigue the ear, thereby masking a following low intensity sound.

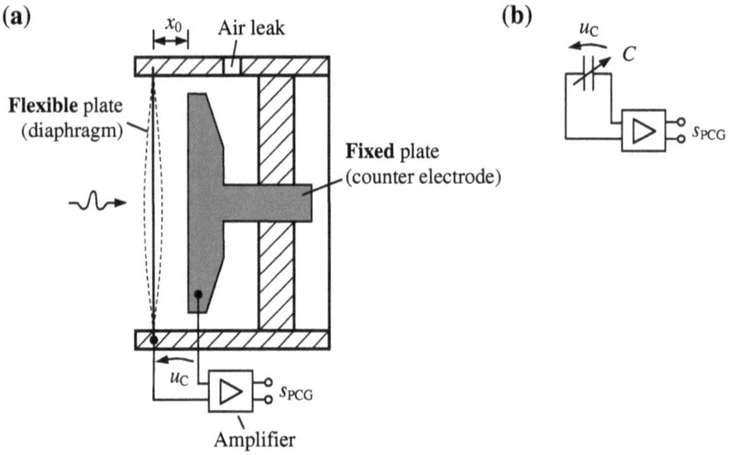

Fig. 4.30 Condenser microphone as an acousto-electric converter. (**a**) The principle set-up. (**b**) The corresponding electrical circuit model

pressure to stretch the skin beneath the diaphragm—the skin operating as a damped *natural diaphragm* (Footnote 52)—increases with decreasing r. Likewise, this makes the eigenfrequency f_{01} of the diaphragm of *low r highly susceptible* to a variation of the *application pressure* (Rappaport and Sprague 1941). Similarly, thick diaphragms and diaphragms with relatively high f_{01} (in the resting state of the diaphragm) are also very susceptible to the variation of the application pressure. However, a pronounced attenuation of low frequencies also causes a pronounced attenuation of the *total sound intensity* passing through the diaphragm (Rappaport and Sprague 1941), which is a disadvantageous issue in the sound auscultation.

In analogy, the resonance curve of the bell, namely, of *Helmholtz resonator*, moves towards higher frequencies with decreasing volume V (4.18), facilitating *absorption and transmission* of high frequency body sounds; compare Fig. 4.29. The width of the resonance curve remains almost unchanged; likewise, the *quality factor* (Footnote 51) of Helmholtz resonator—with respect to the sound absorption—increases with decreasing V.

As shown in Rappaport and Sprague (1941), especially below 200 Hz the bell shows increasing transmission efficiency with increasing V and increasing (input) diameter of the bell. The authors in Ertel et al. (1966b) demonstrate that the response of the bell to the frequencies above approximately 100 Hz strongly deteriorates with the *bell getting shallow*, whereas the response to lower frequencies below 100 Hz remains almost unchanged. It is likely that, while the bell gets shallow, the resonance frequency f_{HR} moves to a higher value because of decreasing V, the *quality factor* of the resonator decreases, and thus the *resonance curve* widens due to increasing frictional losses in the flattened resonator. As reported in Hollins (1971), Ertel et al. (1971), *deep trumpet-shaped bells* provide amplification of body sounds at low frequencies (below approximately 100 Hz) and

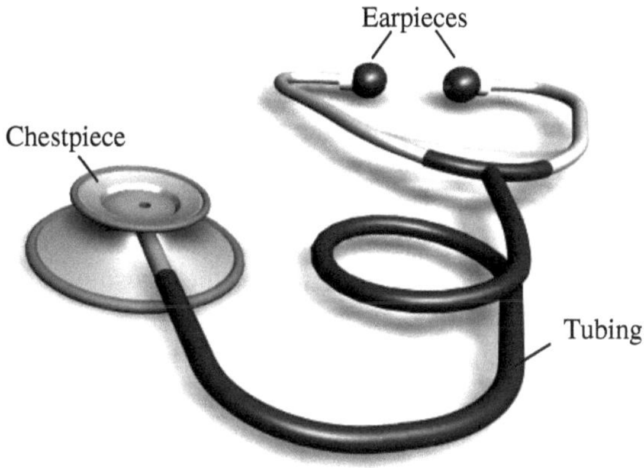

Fig. 4.31 Typical binaural stethoscope including a single chestpiece, flexible tubing, and two earpieces

may also provide amplification at high frequencies (above 100 Hz). In contrast, *shallow bells* provide amplification at low frequencies and predominantly attenuate high frequencies; compare Fig. 4.32. It should be stressed that the latter *amplification* was reported for the combination of the *bell* with a *tubing* connecting to *earpieces* (as intrinsic parts of the standard stethoscope, Fig. 4.31), i.e., the amplification means higher sound intensity at the output of earpieces in comparison with the input of the bell. That is, the reported amplification was also significantly influenced by the sound *transmission pathway in the tubing*, as discussed in Sect. 4.2.1.3. In addition, *bell materials* of moderate density favour the auscultation of low sound frequencies (Hollins 1971).

Furthermore, the larger is the *entrance area of the bell* for the incoming body sounds, the more efficient is the bell at lower frequencies of sounds (Rappaport and Sprague 1941). This behaviour can be attributed to the circular skin region encompassed by the rim of the bell, which was previously referred to as the *natural diaphragm*. That is, increasing r of the bell (Fig. 4.28a) yields lower f_{01} of this natural diaphragm (4.17) and thus manifests in higher efficiency of a widely open bell for the recording of low frequency body sounds.

After the discussion of the resonance behaviour of the chestpiece, the *transmission efficiency of the bell* in terms of the resulting *magnitude of acoustic waves* within the bell should be addressed (Hollins 1971; Ertel et al. 1966b; Rappaport and Sprague 1941). Provided a constant excitation of the bell by body sounds at the skin surface, the magnitude of sound in the bell is

- proportional to the *entrance area of the bell*, i.e., to the surface area ($= \pi \cdot r^2$) of the diaphragm, and
- inversely proportional to the *volume* V to a certain extent (compare Fig. 4.28a).

In fact, the *first criterion* results from the effort to gather body sounds from the largest possible area on the skin. However, increasing surface area is limited by the problem of obtaining a good acoustic seal at the (usually curved) surface of the chest wall. The *second criterion* related to V follows from the consideration that body sounds are to be transmitted from the large-area diaphragm (for sound collection) to the small-area output channel (for sound recording); i.e., the inequality $r \gg r_O$ applies (Fig. 4.28a). Thus an infinitely small V would produce a maximum variation in the sound pressure (or maximum sound intensity) within the bell because the *spring constant of the air* (such as the effective stiffness of the air) in the bell is inversely proportional to V. As discussed in Hollins (1971), Rappaport and Sprague (1941), *small internal volumes* of transmission pathways increase the sound magnitude and thus favour the transmission efficiency. In other words, the vibration magnitude of the air in the region of the diaphragm is much less than the vibration magnitude in the region of the output channel, given the inequality $r \gg r_O$ from above (Fig. 4.28a). However, too small values of V yield high *frictional resistance* during oscillatory movements of the enclosed air and thus introduce additional transmission losses within the bell (compare Sect. 4.2.1.3).

It is interesting to conclude that the above criteria are optimally met if the inner cavity of the bell has a *shape of a funnel* which shows simultaneously a *minimum* V and a *maximum* $\pi \cdot r^2$. In general, other physical factors also contribute to the transmission losses in the bell such as the *surface hardness* of the bell cavity and *air leaks* between the bell and the clamped diaphragm (Abella et al. 1992; Rappaport and Sprague 1941), as discussed later.

Diaphragm and Bell

The *interplay between the diaphragm and bell* should be discussed in terms of the transmission acoustics of the chestpiece. In general, the diaphragm attenuates the *entire transmission pattern* of the chestpiece when compared to the same chestpiece with removed diaphragm (Ertel et al. 1966b); compare Fig. 4.32. In particular, the *bell alone* favours the transmission of *low frequency sounds*, whereas these sounds are suppressed by the *diaphragm* which favours the transmission of *high frequency sounds*. Obviously, the difference between the transmission acoustics of the diaphragm and bell results from the different resonance characteristics of the diaphragm and bell; in particular, it results from the differing values of f_{01} (4.17) and f_{HR} (4.18) and from the differing quality factors; compare Fig. 4.29.

Rigid *diaphragms* attenuate lower frequencies and thereby accentuate faint, high frequency murmurs (Ertel et al. 1971; Rappaport and Sprague 1941). The diaphragms act as *high-pass filters* within the relevant range of body sounds up to 2 kHz and *unmask* high pitched sounds (Footnote 54). As discussed in Welsby and Earis (2001), Hollins (1971), Ertel et al. (1966b), Rappaport and Sprague (1941), the presence of the diaphragm reduces the magnitude of *masking low frequency sounds* (Footnote 54) and thus allows better characterization and *identification of high frequency sounds*, i.e., of informative and usually pathologic body sounds (Footnote 36). However, as already mentioned, the deliberate unmasking can not be attained without lowering the sensitivity of the chestpiece throughout its entire frequency range (Rappaport and Sprague 1941), as can also be observed in Fig. 4.32.

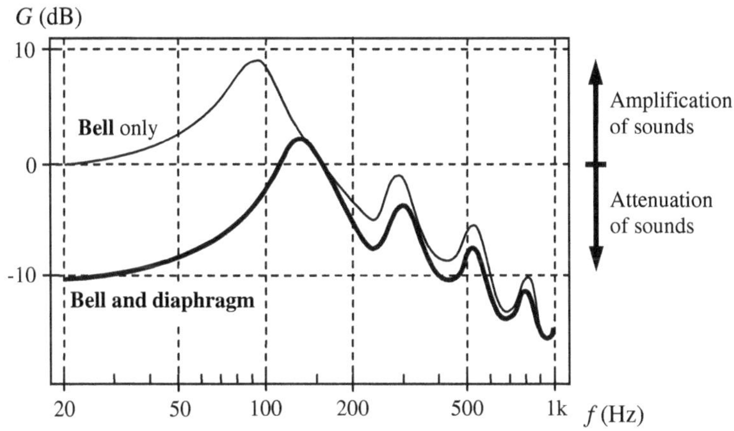

Fig. 4.32 Acoustic transfer function *G* of the stethoscope which includes the chestpiece (with a shallow bell), rubber tubes, and earpieces (Fig. 4.31). The response is defined as the ratio of the sound pressure at the earpiece output to the sound pressure at the chestpiece entrance. The depicted experimental data are taken from Ertel et al. (1966b)

The authors in Hollins (1971), Rappaport and Sprague (1941) show that the *bell alone* is superior to the *bell with the diaphragm* as sound transmitter in the frequency range up to about 400 Hz, whereas the superiority is best at about 300 Hz (Welsby et al. 2003). For sound frequencies above 400 Hz, the reverse is true. The sound *amplification* of the chestpiece (with the diaphragm) was also reported in Kaniusas (2006). According to Abella et al. (1992), in the low frequency range 37–112 Hz typical *chestpieces with the bell alone amplify* body sounds by about 2 up to 12 dB,[55] whereby *chestpieces with the diaphragm attenuate* the sound transmission in most cases by about 5 up to 20 dB; compare Fig. 4.32. In the high frequency range 125–1,000 Hz, *all chestpieces* (with and without the diaphragm) *attenuate* by about 10 dB on average and the attenuation increases with the sound frequency. It should be stressed that the reported quantitative data from Abella et al. (1992) is related to the combination of the chestpiece, tubing, and earpieces (Fig. 4.31), including a strong influence of the tubing on the transmission acoustics (Sect. 4.2.1.3).

Although the *diaphragm* filters out only relatively low frequencies, it is the transmission pattern of the *bell* which predominantly determines whether there is sufficient sound level for high frequencies of sound to be audible (Rappaport and Sprague 1941). Admittedly, there is a trade-off between

[55] Surprisingly, the original *Laennec's stethoscope* (Fig. 1.9), i.e., a simple wooden cylinder, has been shown to amplify body sounds by about 18 dB at the sound frequency of 200 Hz, as noted in Ertel et al. (1966b), Hollins (1971). Surprisingly, this *high amplification* value is comparable with those of modern chestpieces, which makes it difficult to espouse an optimistic view of continuing acoustical improvement of the chestpiece over nearly two centuries (Sect. 1.2.1) aside from the convenience and aesthetic of the modern chestpiece (Fig. 4.31).

- the *detection sensitivity* of body sounds, i.e., the sensitivity is best *with the bell only* and thus without the diaphragm-related losses of sound, and, on the other hand,
- the *characterization* of body sounds (or their identification), i.e., the characterization is best *with the diaphragm* in place.

It follows from the above that physical properties of the chestpiece influence strongly the transmission properties of the different *body sounds* (Sect. 4.1.1). The auscultation of *heart sounds* requires a bell-shaped, open-ended chestpiece (without the diaphragm) which favours the transmission of low frequency body sounds. In contrast, the auscultation of *breathing sounds* such as lung sounds requires a semi-rigid diaphragm covering the flat chestpiece and favouring the identification of high frequency sounds. In the latter case, the *low intensity and high frequency breathing sounds* are unmasked in view of the *high intensity and low frequency heart sounds* (Sect. 4.2.2.3); likewise, high frequency breathing sounds appear to be amplified. In addition, *heart sounds*, i.e., low frequency components of body sounds, can be suppressed by applying *firm application pressure* on the chestpiece bell (Footnote 52). Similarly, *breathing sounds*, i.e., high frequency sound components, are more readily heard with the firmly applied bell or with the diaphragm in use.

Air Leaks

Finally, it should be noted that *air leaks* between

- the chest wall and the rim of the bell or, in analogy, the chest wall and the diaphragm clamped by the bell,
- the bell and the clamped diaphragm, as well as
- the output channel and the microphone

increase not only the *transmission losses* along the sound propagation pathway but also *reshape the acoustic transfer function* of the chestpiece. Likewise, *leak tightness* (good sealing) favours the transmission efficiency of the chestpiece and, in particular, favours the *auscultation of low frequency body sounds*. That is, air leaks act as *high-pass filters*[56] (Rappaport and Sprague 1941) which reduce the amount of low frequency sound components. In addition, air leaks deteriorate the immunity of the chestpiece to external *sound interferences*.

---

[56] An *air leak*—if there is one in the chestpiece—behaves as a *high-pass filter*. In general, the larger is the air leak the more balanced are the sound pressures inside and outside of the chestpiece because of leaking air down the pressure gradient. While the *frequency of sound* gets lower, there is more time for the air to leak out or in, which equalises the latter pressures to a larger extent. As a result, the transmission of low frequency sounds deteriorates provided that an air leak is present. Conclusively, the air leak acts as a high-pass filter. It is interesting to note that some stethoscopes had even an *adjustable leak valve*, i.e., an adjustable high-pass filter. This valve was used to regulate the amount of (low frequency) *heart sounds* reaching the output of the stethoscope (Rappaport and Sprague 1941).

4.2.1.2 Microphone

The *microphone* terminates the *output channel of the chestpiece* bell, as shown in Fig. 4.28a. The microphone is an *acousto-electric converter*. It converts the *sound pressure* variation at the end of the output channel, i.e., at the open neck of Helmholtz resonator (Fig. 4.28b), into an electric signal. This signal serves as the input to a subsequent *signal amplifier*. The output of the amplifier is the *acoustic biosignal phonocardiogram* which instantaneous amplitude is proportional to the instantaneous sound pressure.

The microphone—as illustrated in Fig. 4.30—is usually realised as a *capacitor*, known as the *condenser microphone*. It comprises a metallic diaphragm, i.e., a *flexible plate*, spaced at a short distance from a parallel (massive) *fixed plate*. Both plates act as electrodes of the capacitor. The *sound pressure* variation at the flexible plate bends this plate, which changes the microphone capacitance C.[57] Likewise, the phase front of the acoustic pressure wave—but not the phase front of the wave of the sound particle velocity—modulates the momentary size of C. Since the electric charge accumulated on the capacitor plates is nearly constant, the *voltage* u_C across the capacitor (Fig. 4.30) varies instantaneously in response to the change of C,[58]

---

[57] The *capacitance* C of the *condenser microphone* from Fig. 4.30b can be approximated as

$$C = \frac{\varepsilon \cdot A}{x_0 + \Delta x},$$

where ε is the dielectric permittivity of the air between the plates, A the cross sectional area of the fixed plate, x_0 the distance between the plates, and Δx the change of this distance due to the sound pressure wave; compare Sect. 6. The above *approximation* holds only if the inequality $\Delta x \ll x_0$ applies and Δx is constant over the entire A. It should be noted that the size of A is nearly equal to the cross sectional area of the flexible plate and that of the output channel, see Fig. 4.28a. For instance, if a *harmonic oscillation* of the flexible plate is assumed (Fig. 4.30a) in response to the sound pressure variation at the flexible plate then $\Delta x = X \cdot \cos(\omega t)$ with X as the amplitude of this oscillation.

[58] The *voltage* u_C across the *capacitor* which carries the *electric charge* Q is given by

$$u_C = \frac{Q}{C} = \frac{Q}{\varepsilon \cdot A} \cdot (x_0 + \Delta x);$$

compare Footnotes 33 in Sect. 2 and 57. For the sound frequencies

$$f \gg \frac{1}{2\pi \cdot R \cdot C},$$

the level of Q *remains nearly constant*. Here R denotes the resistance of the RC circuit within the signal amplifier operating the condenser microphone (Fig. 4.30b). In other words, if the time constant $R \cdot C$ is much larger than the oscillation period of the sound, i.e., the *operating circuit* is too inert to follow instantaneously the changes in the sound pressure, then the current $\mathrm{d}Q/\mathrm{d}t$ through the capacitor is almost zero and Q *changes* are negligible. The latter condition is fulfilled for typical frequencies of *body sounds*, which yields that the voltage u_C is a function of Δx only, to be more precise, Δu_C is (approximately) linearly dependent on Δx.

i.e., possible discharge of the capacitor is much slower than the oscillatory change of C with the sound frequency. In qualitative terms, increasing sound pressure at the flexible plate decreases the distance between the plates, increases the size of C (Footnote 57) and thus decreases the instantaneous level of u_C (Footnote 58).

The *waveform of* u_C follows the *sound pressure* waveform at the flexible plate with a high *waveform and phase fidelity*, provided that the corresponding change of C is relatively small. In fact, the voltage u_C goes above and below the bias voltage (= u_C at $\Delta x = 0$, Footnote 58) in synchrony with the momentary sound pressure. The condenser microphone typically shows an *acoustic transfer function* which is practically linear in the frequency range 20–20,000 Hz (within 3 dB range). *Air leaks* in the microphone casing—as shown in Fig. 4.30a—are deliberately used to adjust the acoustic transfer function; compare Footnote 56.

The microphone within the chestpiece applied on the chest wall (Fig. 4.28a), which will be referred to as the *skin microphone*, should be put into perspective with an ambient *room microphone*, i.e., an air-coupled condenser microphone for the registration of body sounds. Obviously, there are strong differences in the *sound coupling*. In order to reach the *skin microphone*, body sounds must pass through the lung parenchyma (as applicable e.g., for *lung sounds*), solid tissues of the mediastinum (applicable for *heart sounds*), pharynx, tracheal and bronchial airways (applicable for *snoring sounds*), and lastly cross the chest skin. Specific filtering effects apply on the transmitted body sounds, as described in Sect. 4.1.2.1.

In contrast, the *room microphone* registers only those body sounds which can be heard by the naked human ear and, to be more precise, which have passed *airway-bound routes* (Fig. 4.23), the pharynx and the nose or mouth cavity (Fig. 4.10). Conclusively, the *room microphone* is applicable for the auscultation of only *lung sounds and snoring sounds*. *Heart sounds* (i.e., low frequency sounds) are not expected to enter respiratory airways—with an aperture towards the room microphone—during their propagation through solid tissue of the mediastinum (Fig. 4.23). Only sounds leaving the pharynx out of the nose or mouth can reach the room microphone; compare also with the effect of small openings on the sound diffraction (see section “Inhomogeneity Effects” in Sect. 4.1.2.2). The latter airy pathway has also a major filtering effect on the transmitted body sounds, amplifying some frequency components and damping others (Fig. 4.24).

Provided the increasing *sound absorption coefficient* with increasing sound *frequency* (see section “Volume Effects” in Sect. 4.1.2.2) and the dependence of the sound *propagation pathway* on the frequency (see section “Specific Issues” in Sect. 4.1.2.1), it can be concluded that

- *high frequency sounds* tend to dominate in the *room microphone* while
- *low frequency sounds* tend to dominate in the *skin microphone*.

In particular, recording of *heart sounds* requires the *skin microphone* (within the chestpiece). In contrast, *lung and snoring sounds* can be auscultated by both the *room microphone* and the *skin microphone*. The sound sources of breathing sounds (i.e., high frequency sounds) reside in relatively large airways, whereas the associated sounds induced tend to propagate along (branched) airways outwards from the body towards both microphones.

However, the room microphone is strongly influenced by *environmental noise* and *position of the subject* (namely, position of the subject's airways) relative to the room microphone. In contrast, the skin microphone within the chestpiece (affixed on the chest wall) ideally receives only body sounds emitted by the skin. It is assumed that the chestpiece does not move with respect to the skin, i.e., acoustical *movement artefacts* can be neglected, and there is a *sound-proof isolation* of the skin microphone from ambient noise.

4.2.1.3 Stethoscope

The *transmission acoustics of the stethoscope* will be discussed, whose basic component is the chestpiece (Sect. 4.2.1.1). In fact, the stethoscope is acoustically a unique device, in that it accounts for acoustical properties of the human ear[59] (Rappaport and Sprague 1941; Ertel et al. 1966b). A *typical stethoscope*—with a modern realisation illustrated in Fig. 4.31 and its precursor, the original Laennec' stethoscope, illustrated in Fig. 1.9—includes

- the bell-type *chestpiece* for *coupling and amplification* of body sounds (compare Fig. 4.27),
- the (rubber) *tubing* as an intermediate connecting element for sound *transmission*, and
- *earpieces* for sound *delivery* into human ears.

As demonstrated in Fig. 4.32, the transmission and filtering pattern of the stethoscope in the frequency domain, namely, the *acoustic transfer function*, does not show a linear sound amplification but almost *regular amplification peaks* (or weak attenuation peaks) alternating with troughs of attenuation (or troughs of strong attenuation). In particular, the amplitudes of the dominant amplification peaks vary with the sound frequency, which indicates the influence of the transmission acoustics of the *chestpiece*, i.e., the bell only or the bell with the clamped diaphragm (Fig. 4.29). Such acoustic transfer functions have been discussed and published in Ertel et al. (1966a, b, 1971) with their critical review in Hollins (1971) and a follow-up study in Abella et al. (1992).

It can be observed in Fig. 4.32 that the *bell with the diaphragm* strongly attenuates low frequency sounds below more than 100 Hz, contrary to the *bell without the diaphragm*. In the latter case, the first (primary) amplification peak

[59] For instance, *human ears* alter the performance of the whole *stethoscope* because earpieces of the stethoscope are terminated with the acoustical impedance of ears and, on the other hand, the latter impedance varies with the sound frequency. As a practical consequence, an *artificial ear*, i.e., a mechanical ear analog with the acoustics of human ears, should be incorporated into experimental systems for the assessment of the objective acoustic transfer functions of stethoscopes (Ertel et al. 1966a, b); compare Fig. 4.32.

occurs at about 100 Hz, the second one occurs at about 300 Hz, whereas other peaks follow at a multiple of about 200 Hz. The presence of the diaphragm lowers the first amplification peak (from about +9 to +2 dB) and shifts this peak to higher frequencies (from about 90 to 130 Hz).

The *undulating pattern* suggests that the *standing waves* within the tubing contribute significantly to the acoustic transfer function of the stethoscope; compare section "Specific Issues" in Sect. 4.1.2.1. In fact, the acoustic propagation lumen within the *tubing* can be approximated as an *open resonating cavity*, compare with Fig. 4.24a. This resonating cavity is *acoustically sealed* at its end with the chest-piece—in analogy with the closed end in Fig. 4.24a—and is *acoustically open* at its end with earpieces—in analogy with the mouth opening in Fig. 4.24a. Consequently, the standing waves arise only when the axial extension of the tubing matches $\lambda/4$, $3 \cdot \lambda/4$, or $5 \cdot \lambda/4$ of sound waves passing through the tubing. A node of the sound pressure and a node of the sound particle velocity result at the opposite ends of the tubing lumen. The resulting *non-harmonic eigenfrequencies* can be calculated according to (4.9), i.e., the sound frequencies at which *body sounds*—more precisely, components of body sounds—are best transmitted through the *tubing* (and even amplified within it) from the chestpiece to earpieces (Fig. 4.31).

In *quantitative terms*, the observed peaks in Fig. 4.32 for the bell without the diaphragm yield—in the context of (4.9)—an axial *extension of the open resonating cavity* of about 86 cm. In particular, the frequencies $f_F^1 = 100$ Hz and $f_F^2 = 300$ Hz are assumed to correspond with $\lambda/4$ and $3 \cdot \lambda/4$ extensions of the cavity, respectively. This estimated axial extension is the approximate axial length of the typical tubing, which strongly indicates that the acoustic resonating lumen is the lumen of the tubing.

In general, the circular *skin* region beneath the diaphragm, already referred to as the *natural diaphragm* (Sect. 4.2.1.1), should be taken into consideration regarding the *acoustic transfer function* of the chestpiece and, consequently, of the whole stethoscope (Rappaport and Sprague 1941; Hollins 1971). For instance, the *primary amplification peak* in the frequency range of about 100 Hz—as illustrated in Fig. 4.32—can not be expected to arise in praxis, since the depicted experimental data from Ertel et al. (1966b) do not consider this natural diaphragm. In fact, the *eigenfrequency* f_{01} (4.17) of the natural diaphragm can be even higher than that of the artificial diaphragm clamped by the bell, despite the lightest application of the chestpiece on the skin (Footnote 52). Likewise, the *resonance curve* of the natural diaphragm is located at a relatively high f_{01} and thus damps the relatively low sound frequencies (at around 100 Hz) which are normally amplified by the bell only. However, it is very difficult to asses quantitatively the behaviour of the natural diaphragm in an experimental way.

As already discussed in Sect. 4.2.1.1, the length and inner diameter of the *tubing* (connecting to earpieces, Fig. 4.31) should be as small as possible because the *inner volume* of the tubing is inversely proportional to the magnitude of the acoustic pressure wave within the tubing. However, the smaller is the inner diameter of the tubing, the higher is the local *frictional resistance*—to which the oscillating air column in the tubing is subjected—and thus the lower is the resulting magnitude of the *pressure wave*. In other words, a *compromise* should be reached between the inner volume (inner diameter) and the frictional resistance of the tubing. Authors in Rappaport and Sprague (1941), Hollins (1971), Abella et al. (1992) prove experimentally that a *longer tubing* and a *smaller diameter* tend to yield *greater sound reduction*. For instance, an increase in the tubing length from about 8 to 66 cm contributes about 15 dB attenuation of the sound pressure at the sound frequency of 200 Hz. The above effect of the varying tubing length dominates above 100 Hz, whereas the tubing length minimally affects the sound attenuation at frequencies below 100 Hz (Rappaport and Sprague 1941).

Furthermore, the *wall of the tubing* should be sufficiently rigid and the *interior surface* sufficiently smooth to attain maximum transmission efficiency of the tubing (Abella et al. 1992; Rappaport and Sprague 1941). This is because any wall motion and increased frictional resistance reduce the effective variation of the sound pressure in the tubing. In addition, *air leaks* in the tubing also contribute to the sound attenuation (Abella et al. 1992), as discussed in Footnote 56.

4.2.2 Registration of Body Sounds

As soon as *body sounds* have been coupled into the *chestpiece* (Sect. 4.2.1.1) and converted into an acoustic biosignal *phonocardiogram* (Sect. 4.2.1.2), body sounds are available for signal analysis and *diagnostic purposes*. That is, the phonocardiogram reflects inner body sounds which, in principle, are *mechanical waves* echoing the *mechanical function* of

- *cardiac system* and
- *respiratory system*.

The following sections will demonstrate *experimental phonocardiograms* recorded from the *heart region on the chest*, as illustrated in Fig. 4.33. *Numerous physiological parameters* and events related to the cardiac and respiratory systems will be derived out of phonocardiograms. Likewise, *multiparametric data* (Sect. 1.4) will be derived from a *single* acoustical sensing device. In particular, the *sensing*

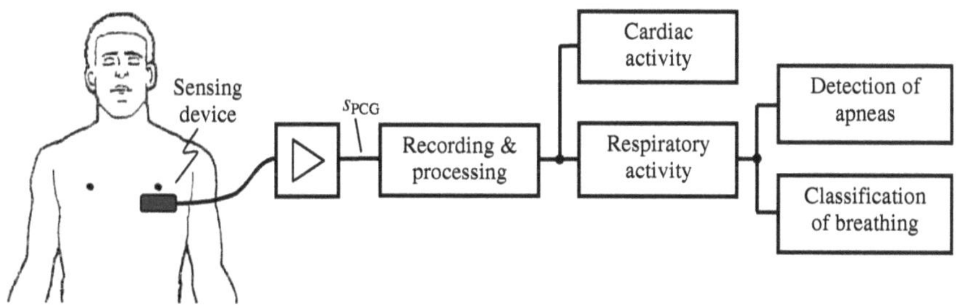

Fig. 4.33 Registration of body sounds on the chest wall and their multiparametric processing in order to extract various cardiorespiratory parameters; compare Fig. 4.1

device, i.e., the chestpiece with a microphone at its output (Fig. 4.28a), was affixed to the chest skin with a double-sided adhesive tape.

The *dynamic nature* of body sounds reveals cardiorespiratory activity. It will be shown that *different frequency ranges* of the different *body sounds* are highly useful for the multiparametric evaluation. For the *chest region*, as summarized in Sect. 4.1.1.5,

- *heart sounds* reside in the frequency range up to 100 Hz,
- vesicular *lung sounds* in the range 100–500 Hz,
- *normal snoring sounds* in the range 100–800 Hz, and
- *obstructive snoring sounds* in the range 100–2,000 Hz.

In particular, *signal power* of the phonocardiogram—compare Footnote 193 in Sect. 3—will be estimated in the different frequency ranges to uncover cardiac and breathing activity. That is,

- signal power P_L in the low frequency range up to 100 Hz, accounting for *heart sounds* and thus for *cardiac activity*, will be used; as well as
- signal power P_W in the wide frequency range 100–2,000 Hz, accounting for vesicular *lung sounds* and (normal and obstructive) *snoring sounds* and thus for the *breathing activity*.

Besides the use of the signal power, there are many other (direct and indirect) methods to extract cardiac and breathing activity out of the mixed body sounds (Sect. 3.2.1) emanating from the depths of the body (Fig. 4.1).

4.2.2.1 Cardiac Activity

Heart sounds arise in the course of cardiac activity, as discussed in Sect. 4.1.1.1 and exemplified in Figs. 4.5, 4.9 and 4.13. An important (vital) physiological parameter of cardiac activity is the heart rate f_C (Sect. 3.1.1).

Figure 4.34 demonstrates the *registration of* f_C out of heart sounds during an obstructive *sleep apnea*. In the *time domain* (Fig. 4.34a), heart sounds can clearly be distinguished. Obstructive snoring sounds—surrounding the obstructive sleep apnea (Sect. 4.1.1.4)—yield larger deflection amplitudes than heart sounds (compare Fig. 4.14). In the *frequency domain*, the *signal power* P_L was estimated, which time course is depicted in Fig. 4.34b. The oscillation of P_L clearly follows the periodic occurrence of both heart sounds as a unit,[60] as expected from Sect. 4.1.1.1.

[60] It should be noted that the *course of* P_L from Fig. 4.34b does not follow the individual *heart sounds*, i.e., the first or second heart sound, but rather the assembly of both heart sounds as a unit. This is because time intervals for P_L estimation (of 256 ms duration, Fig. 4.34) are larger than individual durations of heart sounds (usually < 140 ms, Sect. 4.1.1.1). Consequently, the instantaneous level of P_L is a sliding average over both heart sounds.

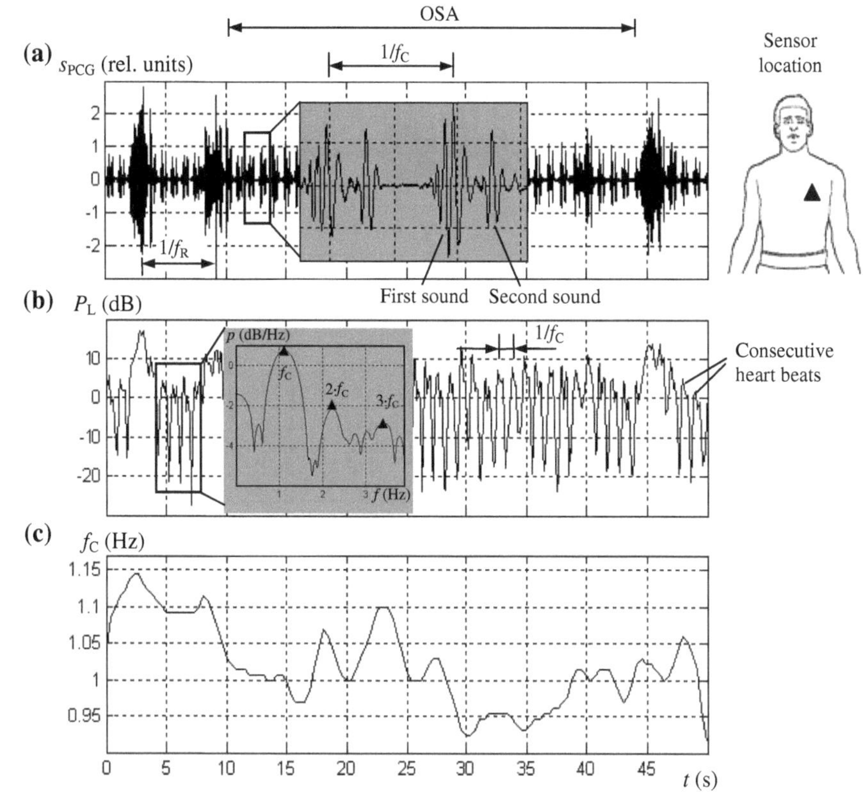

Fig. 4.34 Assessment of cardiac activity by body sounds in the course of an obstructive sleep apnea (OSA). (**a**) Acoustic biosignal phonocardiogram s_{PCG} from the heart region on the chest. (**b**) Signal power P_L of s_{PCG} in the low frequency range up to 100 Hz (calculated for time intervals of 0.256 s duration with 50 % overlap). (**c**) The heart rate f_C derived from the time course of P_L (calculated for time intervals of 4 s duration with 90 % overlap) using signal processing methods in the frequency domain (Footnote 61). The power spectral density p of P_L is depicted in (**b**) for the time interval 3.8–7.8 s, demonstrating multiple peaks at f_C, $2{\cdot}f_C$, and $3{\cdot}f_C$. The respiratory rate f_R is also indicated in (**a**)

Thus, the oscillation rate of P_L is actually the heart rate f_C which estimated time course[61] is shown in Fig. 4.34c. It can be observed that f_C temporarily decreases

[61] In general, the heart rate f_C *can be estimated* from the *time course of* P_L in different ways:

- In the *time domain*, detection of peaks in P_L (or, alternatively, zero crossings in P_L) could be performed, whereas the difference between the corresponding neighbouring timestamps of peaks (or zero crossings) yields the *instantaneous level* of $1/f_C$; compare Figs. 5.31 and 5.33.
- In the *frequency domain*, peaks in the *power spectral density* of P_L could be detected, especially those peaks which reside at multiple frequencies. These multiple frequencies would most likely correspond to *multiple harmonics* of f_C, i.e., f_C, $2 \cdot f_C$, ... $k \cdot f_C$ with k as the integer index. Since a time interval (window) of P_L is used for the calculation of the power spectral

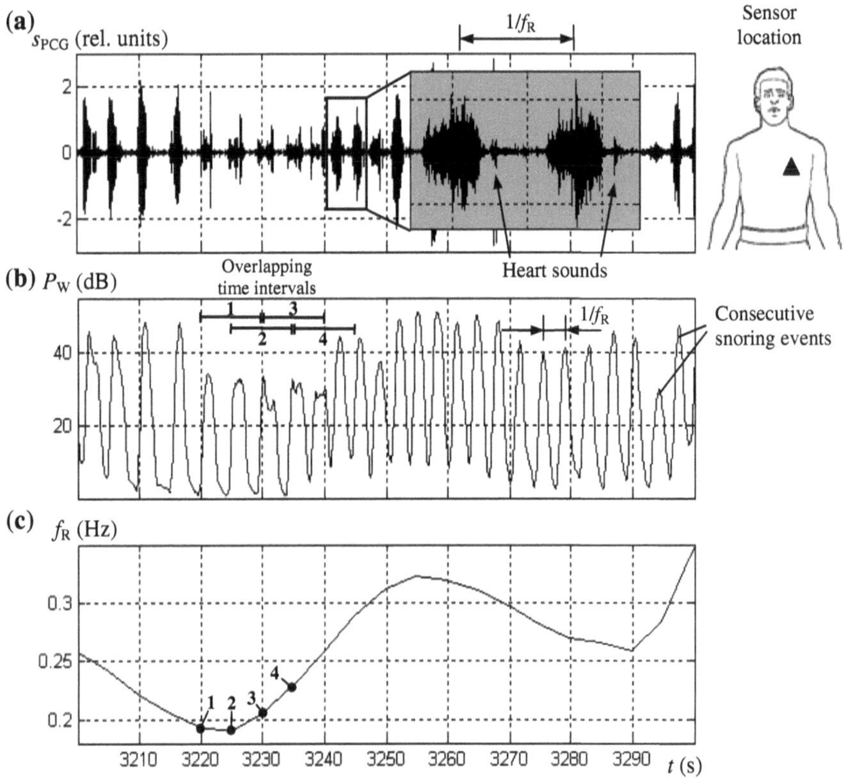

Fig. 4.35 Assessment of respiratory activity by body sounds in the course of obstructive snoring. (**a**) Acoustic biosignal phonocardiogram s_{PCG} from the heart region on the chest. (**b**) Signal power P_W of s_{PCG} in the wide frequency range 100–2,000 Hz (calculated for time intervals of 0.256 s duration with 50 % overlap). (**c**) The respiratory rate f_R derived from the time course of P_W (calculated for time intervals of 10 s duration with 50 % overlap) using signal processing methods in the frequency domain (Footnote 61). Four consecutive time intervals are indicated in (**b**) while the corresponding values of f_R are indicated in (**c**)

during the apnea, which confirms physiological discussions in section "Ceased Respiration" in Sect. 3.2.1.1 and polysomnographic observations in Fig. 3.9.

(Footnote 61 continued)
density and thus the estimation of f_C, this method yields only an *average level* of f_C within this particular time interval (window). The corresponding examples are demonstrated in Figs. 4.34b and 4.35.

4.2.2.2 Respiratory Activity

Breathing sounds, i.e., lung and snoring sounds, arise in the course of respiratory activity, as discussed in Sects. 4.1.1.2–4.1.1.4 and exemplified in Figs. 4.9, 4.13 and 4.14. An important (vital) physiological parameter of respiratory activity is the respiratory rate f_R (Sect. 3.1.2).

Figure 4.35 demonstrates the *registration of* f_R out of breathing sounds during *obstructive snoring*. In the *time domain* (Fig. 4.35a), obstructive snoring sounds can clearly be distinguished (compare Fig. 4.14) with their obvious dominance over heart sounds. In the *frequency domain*, the signal power P_W was estimated, which time course is depicted in Fig. 4.35b. The consecutive snoring events are clearly visible in the oscillating course of P_W, whereas the oscillation rate of P_W is actually the rate f_R. The estimated time course[62] of f_R (Fig. 4.35c) reflects clearly the regions of accelerated and decelerated oscillations of P_W (Fig. 4.35b).

Figure 4.36 demonstrates the course of P_W during two different *sleep apneas*, namely, an obstructive sleep apnea (Fig. 4.36a) and obstructive sleep hypopnea (Fig. 4.36b). In the case of the *obstructive apnea*—this particular apnea is also demonstrated in Fig. 4.15a—large peaks of P_W surround this apnea, whereas these peaks correspond to snoring events. In addition, two minor peaks, i.e., apneic *respiratory efforts* (Sect. 3.1.2), reside directly within this apnea. A *residual cardiac component* can be recognized in P_W (Fig. 4.36a), which oscillates with f_C and indicates minor contributions of heart sounds above 100 Hz to the power P_W. In the case of the *obstructive hypopnea* (Fig. 4.36b)—this particular hypopnea is also demonstrated in Fig. 4.16—the amplitude of P_W peaks is temporarily reduced. This confirms that the intensity of snoring events is diminished during the hypopnea. In fact, *waveform analysis* of P_W allows for the detection of apneas[63] (Kaniusas 2006); compare Fig. 4.33.

---

[62] In analogy with Footnote 61, the respiratory rate f_R *can be estimated* from the time course of P_W in the *time domain* (by detecting peaks or zero crossings) and the *frequency domain* (by detecting multiple harmonics of f_R).

[63] There are numerous methods to *detect sleep apneas* by *acoustical means*. However, strong variability of snoring sounds—or, in general, variability of breathing sounds—within single subjects and even from one breath to another complicates matters (Sect. 4.1.1.3). In fact, apneas are characterised by

- *increased total intensity of breathing sounds* which surround apneas because of deteriorated pharyngeal dynamics (Itasaka et al. 1999; Pasterkamp et al. 1997b). Thus, intensity thresholds and time interval measurements can be applied for *apnea detection* (Brunt et al. 1997). In addition, obstructive snoring shows increased amount of high frequency components (Sect. 4.1.1.3), so that
- *increased partial intensity of breathing sounds* within a limited (specific) frequency range favours *apnea detection* (McCombe et al. 1995; Penzel et al. 1990; Rauscher et al. 1991; Verse et al. 2000).

Usually, an overestimation of the number of apneas is reported, which were detected by acoustical means. The reliability of apnea detection typically increases with the apnea severity (or with the airway obstruction severity) and with respiratory disturbance index (Sect. 3.1.2). Provided that the *waveform of* P_W serves as a basis for apnea detection—as illustrated in Fig. 4.36—

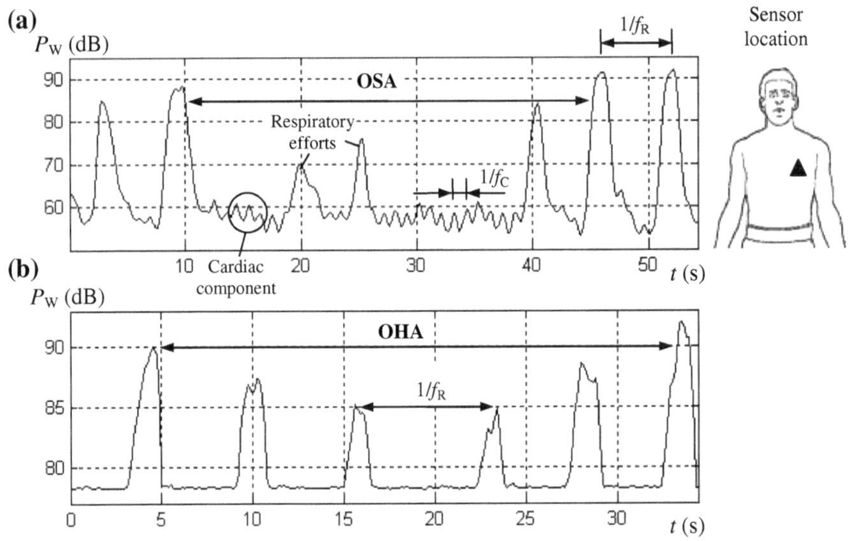

Fig. 4.36 Detection of apneas by the use of body sounds; in particular, by the use of the acoustic biosignal phonocardiogram s_{PCG} from the heart region on the chest. (**a**) Signal power P_W of s_{PCG} in the wide frequency range 100–2,000 Hz (calculated for time intervals of 0.256 s duration with 50 % overlap) in the course of an obstructive sleep apnea (OSA) which is also illustrated in Fig. 4.15a. (**b**) Signal power P_W in the course of an obstructive sleep hypopnea (OHA) which is also illustrated in Fig. 4.16. The respiratory rate f_R and heart rate f_C are indicated

A simplified *classification of breathing* into

- normal breathing,
- normal snoring, and
- obstructive snoring

is highly relevant for diagnosis of respiratory diseases (Sects. 3.1.2 and 4.1.1) and can be addressed by body sounds (Fig. 4.33). However, the above classification is not straightforward because of *strong variability of breathing sounds* (i.e., high dispersion of sound characteristics), *fluent transition* from one breathing type to another, and *missing standards* in the classification of breathing.

In fact, the *frequency range* of body sounds extends *beyond 100 Hz* with the *on set of breathing*; compare the frequency range of heart sounds only with that of breathing sounds (Sect. 4.2.2). Furthermore, the effective frequency range of body sounds widens towards *higher frequencies* while breathing becomes progressively *obstructive*; compare the frequency range of normal snoring sounds with that of

(Footnote 63 continued)
adaptive and time-dependent power *thresholds* can be used. The resulting intervals in combination with time limits facilitate apnea detection (Kaniusas 2006).

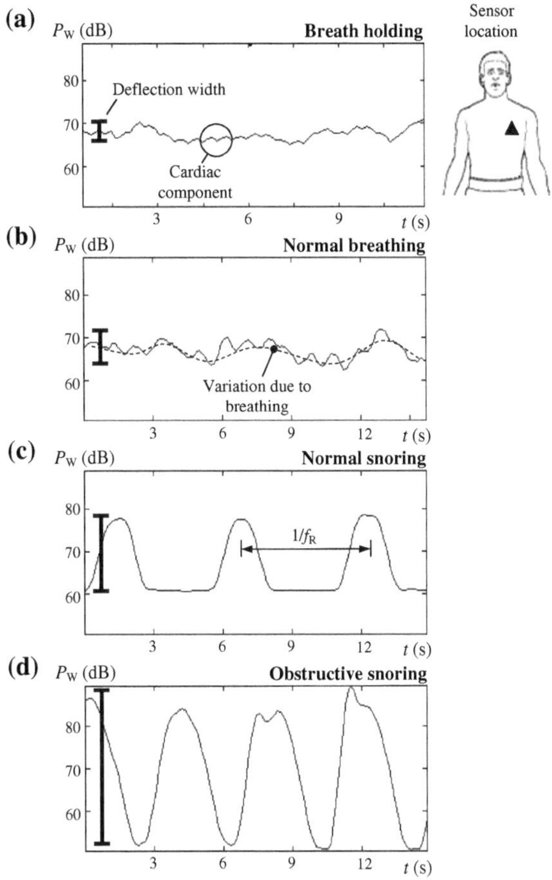

Fig. 4.37 Classification of breathing by the use of body sounds; in particular, by the use of the acoustic biosignal phonocardiogram s_{PCG} from the heart region on the chest. (**a**) Signal power P_W of s_{PCG} in the wide frequency range 100–2,000 Hz (calculated for time intervals of 0.256 s duration with 50 % overlap) during breath holding. (**b**) Signal power P_W during normal breathing. (**c**) Normal snoring. (**d**) Obstructive snoring. The respiratory rate f_R is also indicated

obstructive sounds.[64] Thus, it can be expected that the *signal power* P_W in the wide frequency range 100–2,000 Hz accounts for the different breathing types.

Figure 4.37 demonstrates the course of P_W for *different types of breathing*. In particular, during breath holding (Fig. 4.37a), the *residual cardiac component*

---

[64] The appearance of high frequencies in body sounds is a clear indication for approaching *obstruction*. For instance, as discussed in Kaniusas et al. (2005),

- *normal breathing* shows a predominance of low frequency sound power according to $P_L' > P_M > P_H$,
- *normal snoring* yields $P_L' \geq P_M > P_H$, and
- *obstructive snoring* shows a predominance of high frequency power corresponding to $P_L' \geq P_H \geq P_M$.

Here P_L' is the signal power of the phonocardiogram in the extended low frequency range up to 300 Hz, P_M the signal power in the medium frequency range 300–800 Hz, and P_H the signal power in the high frequency range 800–2,000 Hz; see Sect. 4.1.1.3.

dominates. *Normal breathing* yields a slight superimposed oscillation with f_R (Fig. 4.37b). In contrast, *normal snoring* causes clear peaks (Fig. 4.37c), whereas *obstructive snoring* induces an even higher peak amplitude (Fig. 4.37d). Likewise, the deflection width of P_W increases from breath holding to normal breathing, from normal breathing to normal snoring, and, finally, from normal snoring to obstructive snoring.

Since the deflection width of P_W during *normal breathing* is relatively small (Fig. 4.37b), i.e., lung sounds and silent snoring sounds are almost masked by heart sounds (Figs. 4.9 and 4.13a), it can be expected that here the *registration of* f_R is more challenging than during normal (or obstructive) snoring. Snoring yields large oscillations of P_W (Fig. 4.37c, d) and dominates clearly over heart sounds in the time domain, which facilitates the registration of f_R, as exemplified in Fig. 4.35.

4.2.2.3 Spatial Distribution of Body Sounds

It can be expected that the *intensity of body sounds* is unevenly distributed inside the body as well as on the skin surface. This is of high importance if an *optimal auscultation region* for particular body sounds has to be used.

The *uneven spatial distribution* of body sounds is due to the facts that

- *sources* of body sounds usually *lack spatial symmetry* with respect to the body axis. The special asymmetry is primarily given by the massive mediastinum on the left side of the thorax; compare Fig. 4.20 (Sect. 4.1.1). In addition,
- sound *propagation paths* from sound sources to the skin surface cross *heterogeneous tissues* (such as blood, muscles, bones, air, and lung tissue). This yields a highly heterogeneous sound transmission in the body or, broadly speaking, *varying sound absorption* as a function of space (Table 4.1, Sect. 4.1.2). Furthermore,
- spatial propagation paths in the body even depend on the characteristic *properties of propagating sounds* (such as sound frequency, Fig. 4.23).

The spatial distribution of the intensity of *heart sounds* was investigated by Kompis et al. (1998). The authors demonstrated that *hypothetical sound sources* (estimated sources) of the *first heart sound* are *spatially limited* to the location of the heart itself. Indeed, given the generation mechanisms of heart sounds (from Sect. 4.1.1.1), the sound sources of the first heart sound can be expected to reside in a constricted space within the heart. This is because atrioventricular valves—the origin site of the first heart sound—are located inside the heart and thus the corresponding sound sources are almost *isolated from outside* the heart (compare Fig. 4.3).

In contrast, the *second heart sound* gives rise to a *spatially scattered* pattern of *hypothetical sound sources*. This pattern shows multiple sound sources which are close to the heart region but are spatially separated from each other. It can be explained by the distal location of semilunar valves, the origin site of the second heart sound. Semilunar valves act as output valves which closure induces vibrations of blood *outside the heart*; these vibrations, in turn, contribute to the scattering of multiple sound sources.

The spatial distribution of the intensity of (vesicular) *lung sounds* on the chest wall is consistent with the origin of these sounds (Sect. 4.1.1.2), as proven by many authors (Pasterkamp et al. 1997a; Kompis et al. 1998, 2001; Loudon and Murphy 1984; Fachinger 2003). During inspiration, *hypothetical sound sources* of *inspiratory sounds* are widely *distributed* in terms of *diffuse sources* and reside predominantly in the periphery of the lung (distal airways). In contrast, *expiratory sounds* if heard on the chest wall are rather generated by a *point source* of sound residing in the upper proximal airways. Likewise, expiratory sounds on the chest are similar in quality to tracheobronchial lung sounds.

In addition, the distribution of vesicular *lung sounds* is *asymmetric* on the chest wall (Pasterkamp et al. 1997a, b; Jones et al. 1999; Fachinger 2003). The intensity of surface sounds lateralizes with the *right-over-left dominance* at the anterior upper chest and with the *left-over-right dominance* at the posterior upper chest. This lateralization dominates during *expiration* and for *low frequency sounds*,[65] i.e., for sounds below 300 Hz (Pasterkamp et al. 1997b) or below 600 Hz (Wodicka et al. 1989). In addition, anterior sites show higher intensity of lung sounds than posterior sites.

It is likely that the latter asymmetries are related to

- the *asymmetric* localization of inner *cardiovascular structures* on the left side of the major airways and, on the other hand, related to
- the *asymmetric* morphology and asymmetric distribution of bronchial *airways*.

That is, the preferential coupling of *inspiratory sounds* (i.e., vesicular lung sounds) into the right anterior upper chest could be explained by the massive *mediastinum on the left* side because the mediastinum tends to *attenuate strongly* the propagating sounds towards the left anterior upper chest. In analogy, *expiratory sounds* which propagate towards the left anterior upper chest are also damped by the mediastinum adjacent to the left side of the major airways; expiratory sounds towards the right anterior upper chest are damped to a lesser extent. The effect of the *asymmetric bronchial airways* can be exemplified by the fact that some major left bronchi are directed more posteriorly compared with the right bronchi, which is due to the anterior position and required space of the heart. Obviously, the asymmetric airways favour the left-over-right dominance of the sound intensity at the posterior upper chest because areas of sound generation are closer to the chest wall (or to the skin surface) at posterior sites on the left. It can be observed that the *lateralization* of breathing sounds is compatible with the discussed *concepts* of a

[65] It should be noted that *sound frequencies* which *lateralize* best coincide well with sound frequencies which tend to be *coupled* from the airy respiratory airways into the semi-solid mediastinum (or into the lung parenchyma); compare with the frequency dependent propagation of body sounds in Fig. 4.23 and Section "Specific Issues" in Sect. 4.1.2.1. Thus, it can be expected that *heterogeneous tissues* contribute significantly to the *asymmetric transmission* of vesicular lung sounds at relatively *low sound frequencies*. At relatively *high frequencies*, the asymmetry of the sound transmission is weaker because sounds predominantly prefer airway-bound routes and their pathways are more direct, bypassing the damping effect of the *asymmetric mediastinum*.

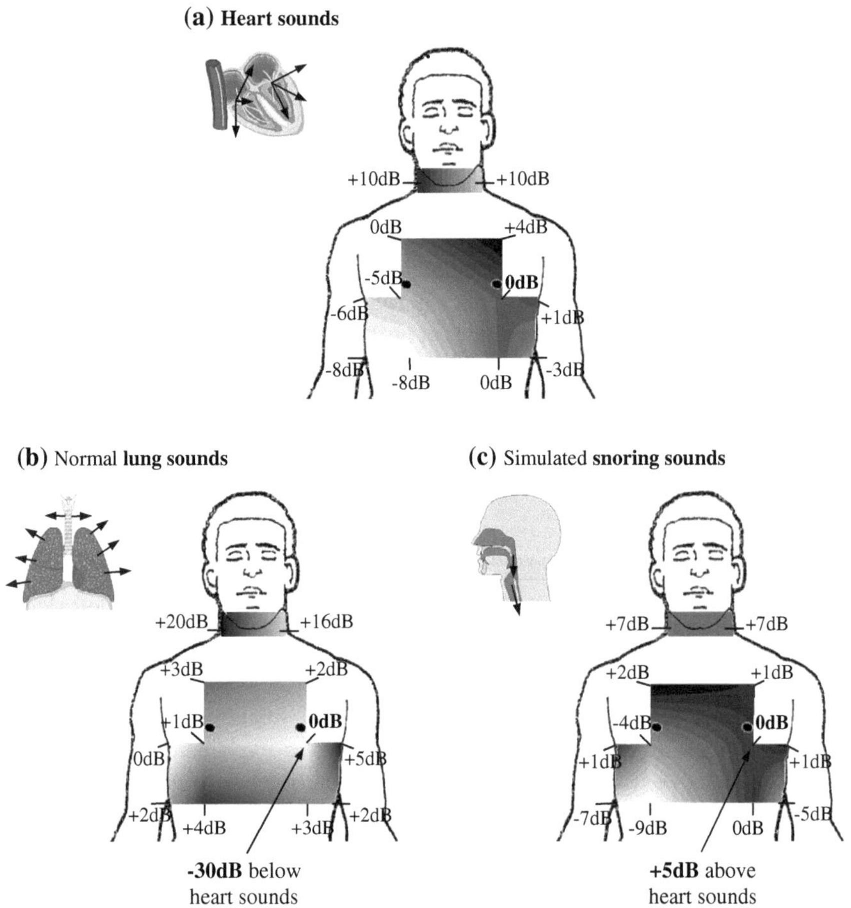

Fig. 4.38 Spatial distributions of body sounds on the upper body of two healthy male subjects. Local variations of the relative sound intensity are shown, whereas the (averaged) intensity level is given in dB in relation to the heart region on the chest (with the local intensity level of 0 dB). Intensity values between the measured points (denoted by explicit numbers) are generated using bilinear interpolation and are indicated through *grey tones*. (**a**) Spatial distribution of heart sounds. (**b**) Distribution of normal lung sounds. In the heart region, the corresponding sound intensity is 30 dB below the local intensity of heart sounds. (**c**) Distribution of simulated snoring sounds. In the heart region, the corresponding sound intensity is 5 dB above the local intensity of heart sounds

central origin of expiratory sounds and a *distributed origin of inspiratory sounds*, when sounds are auscultated on the chest wall (Sect. 4.1.1.2).

The spatial distribution of the intensity of *snoring sounds* on the chest wall can be expected to follow the aforementioned pattern of *expiratory sounds* lateralization because sources of snoring sounds (and those of expiratory breathing sounds, Sect. 4.1.1.2) reside close to the upper airways (Sect. 4.1.1.3). Likewise, this spatial distribution of snoring sounds could be derived from the observed lateralization of *artificial sounds* introduced at the mouth (close to the pharyngeal airway, the origin site of snoring sounds); i.e., derived from the *passive transmission* of sounds from the mouth to the chest wall (Pasterkamp et al. 1997a). Given the above analogies, it can be expected that *snoring sounds* would lateralize with the *right-over-left dominance* at the anterior upper chest. At the posterior upper chest, snoring sounds can be expected to be only slightly louder on the left, which is similar to the lateralisation of vesicular lung sounds due to the asymmetry of bronchial airways. In addition, anterior sites can be expected to show higher intensity of snoring sounds than posterior sites.

As illustrated in Fig. 4.38, intensities of *body sounds* were assessed on ten different *regions on the thorax*; in particular, at around the 3rd, 5th, and 7th intercostal space anterior left and right, respectively, and around the 5th and 7th intercostal space lateral left and right, respectively (Kaniusas et al. 2005). In addition, sound intensities on the neck, collateral to the trachea, were assessed for comparison. Each subfigure in Fig. 4.38 includes data on the *relative sound intensity* in relation to the heart region on the chest (around the 5th intercostal space anterior left). Thus, the relative sound intensity in this heart region amounts to 0 dB.

In comparison with the heart region, the following *tendencies* can be observed within the spatial distributions of Fig. 4.38:

- *Heart sounds* (Fig. 4.38a)—the sound intensity decreases with increasing distance from the heart, i.e., the intensity shows minimum values of about −8 dB in the lower right thorax region. Conversely, the intensity yields a 10 dB maximum at the neck.
- *Lung sounds* (Fig. 4.38b)—the sound intensity shows only slight local differences at the thorax without any evident lateralization, which results from widely distributed sound sources. Strongly enhanced sounds arise at the neck (up to +20 dB) because tracheobronchial sounds (with near sources) are more intense than vesicular sounds (with distant sources); see Sect. 4.1.1.2.
- (Simulated) *snoring sounds* (Fig. 4.38c)—the sound intensity shows a maximum of about +7 dB at the neck because of closely located sources of snoring sounds. The intensity tends to decrease with increasing distance from the neck.

When the discussed distributions of the different body sounds are compared with each other, it can be observed that—in the heart region on the chest—the intensity of *lung sounds* is about 30 dB below that of *heart sounds*. *Snoring sounds* in the heart region are stronger by about 5 dB than local heart sounds. From a practical point of view, it means that *simultaneous auscultation* of a mixture of the different body sounds *disadvantages (faint) lung sounds*; compare Footnotes 53 and 54. The *heart region* on the chest yields a ratio 0 dB: −30 dB: +5 dB (relative intensities of heart: lung:snoring sounds). A more *optimal region* for the simultaneous auscultation of all three body sounds would be the lower right thorax region (around the 7th intercostal space anterior right); it yields a ratio −8 dB: −26 dB: −4 dB or, related to the heart region, 0 dB: −18 dB: +4 dB. Another attractive region would be the neck, whereas its right side yields a ratio +10 dB: −10 dB: +12 dB or 0 dB: −20 dB: +2 dB, respectively. In any case, lung sounds are at a disadvantage.

The *use of spatial distributions* of the different body sounds may lead to advanced diagnostic methods beyond a simple single spot sound auscultation. This has already been proposed for heart sounds (Leong-Kon et al. 1998) and lung sounds (Kompis et al. 2001). For instance, *acoustic images* of a pathologically consolidated lung—by the use of the spatial distribution of the intensity of lung sounds—differ substantially from images of a healthy lung and thus help to localize abnormalities; compare section "Volume Effects" in Sect. 4.1.2.2. Obviously, the *spatial resolution* of acoustic images can not be expected to resolve differences below approximately 2 cm while localising sound sources. This is because the resolution is determined by the size of λ whose lowest reported values are about 2.3 cm (at $v = 23$ m/s and $f = 1$ kHz, see (4.3) and section "General Issues" in Sect. 4.1.2.1).

References

M. Abella, J. Formolo, D.G. Penney, Comparison of the acoustic properties of six popular stethoscopes. J. Acoust. Soc. Am. **91**(4), 2224–2228 (1992)

G. Amit, K. Shukha, N. Gavriely, N. Intrator, Respiratory modulation of heart sound morphology. Am. J. Physiol. Heart Circ. Physiol. **296**(3), 796–805 (2009)

R. Beck, M. Odeh, A. Oliven, N. Gavriely, The acoustic properties of snores. Eur. Respir. J. **8**(12), 2120–2128 (1995)

N. Brooks, G. Leech, A. Leatham, Factors responsible for normal splitting of first heart sound. High-speed echophonocardiographic study of valve movement. Br. Heart J. **42**(6), 695–702 (1979)

D.L. Brunt, K.L. Lichstein, S.L. Noe, R.N. Aguillard, K.W. Lester, Intensity pattern of snoring sounds as a predictor for sleep-disordered breathing. Sleep **20**(12), 1151–1156 (1997)

A. Bulling, F. Castrop, J.D. Agneskirchner, W.A. Ovtscharoff, L.J. Wurzinger, M. Gratzl, *Body Explorer. An Interactive Program on the Cross-Sectional Anatomy of the Visible Human Male* (Springer, Berlin, 1997)

D. Chamier, Unpublished ball pen drawing, Institute of Art and Design, Vienna University of Technology (2014)

F. Cirignota, Classification and definition of respiratory disorders during sleep. Minerva Med. Rev. **95**(3), 177–185 (2004)

F. Dalmay, M.T. Antonini, P. Marquet, R. Menier, Acoustic properties of the normal chest. Eur. Respir. J. **8**(10), 1761–1769 (1995)

J. Earis, Lung sounds. Thorax **47**(9), 671–672 (1992)

K.R. Erikson, F.J. Fry, J.P. Jones, Ultrasound in medicine—a review. IEEE Trans. Sonic Ultrason. **21**(3), 144–170 (1974)

P.Y. Ertel, M. Lawrence, R.K. Brown, A.M. Stern, Stethoscope acoustics I. The doctor and his stethoscope. Circulation **34**(5), 889–898 (1966a)

P.Y. Ertel, M. Lawrence, R.K. Brown, A.M. Stern, Stethoscope acoustics II. Transmission and filtration patterns. Circulation **34**(5), 899–909 (1966b)

P.Y. Ertel, M. Lawrence, W. Song, Stethoscope acoustics and the engineer: concepts and problems. J. Audio Eng. Soc. **19**(3), 182–186 (1971)

P. Fachinger, Computer based analysis of lung sounds in patients with pneumonia—Automatic detection of bronchial breathing by Fast-Fourier-Transformation (in German: Computerbasierte Analyse von Lungengeräuschen bei Patienten mit Pneumonie—Automatische Detektion des Bronchialatmens mit Hilfe der Fast-Fourier-Transformation). Dissertation, Philipps-University Marburg, (2003)

D.C. Giancoli, Physics (in German: Physik). Pearson Studium (2006)

L.J. Hadjileontiadis, S.M. Panas, in *Nonlinear Separation of Crackles and Squawks from Vesicular Sounds Using Third-Order Statistics*. Proceedings of the 18th Annual EMBS International Conference, vol. 5, (1996), pp. 2217–2219
L.J. Hadjileontiadis, S.M. Panas, Adaptive reduction of heart sounds from lung sounds using fourth-order statistics. IEEE Trans. Biomed. Eng. **44**(7), 642–648 (1997a)
L.J. Hadjileontiadis, S.M. Panas, Separation of discontinuous adventitious sounds from vesicular sounds using a wavelet-based filter. IEEE Trans. Biomed. Eng. **44**(12), 1269–1281 (1997b)
W. Hohenhorst, Unpublished image data, Clinic of Otolaryngology, Alfried Krupp Hospital, Germany (2000)
P.J. Hollins, The stethoscope. Some facts and fallacies. Br. J. Hosp. Med. **5**, 509–516 (1971)
K. Ishikawa, T. Tamura, Study of respiratory influence on the intensity of heart sound in normal subjects. Angiology **30**(11), 750–755 (1979)
Y. Itasaka, S. Miyazaki, K. Ishikawa, K. Togawa, Intensity of snoring in patients with sleep-related breathing disorders. Psychiatry Clin. Neurosci. **53**(2), 299–300 (1999)
V.K. Iyer, P.A. Ramamoorthy, Y. Ploysongsang, Autoregressive modeling of lung sounds: characterization of source and transmission. IEEE Trans. Biomed. Eng. **36**(11), 1133–1137 (1989)
A. Jones, R.D. Jones, K. Kwong, Y. Burns, Effect of positioning on recorded lung sound intensities in subjects without pulmonary dysfunction. Phys. Ther. **79**(7), 682–690 (1999)
E. Kaniusas, Multiparametric physiological sensors. Habilitation theses, Vienna University of Technology, 2006
E. Kaniusas, in *Acoustical Signals of Biomechanical Systems*, ed. by C.T. Leondes. Biomechanical Systems Technology, vol. 4 (World Scientific Publishing, Singapore, 2007), pp. 1–44
E. Kaniusas, H. Pfützner, B. Saletu, Acoustical signal properties for cardiac/respiratory activity and apneas. IEEE Trans. Biomed. Eng. **52**(11), 1812–1822 (2005)
T. Koch, S. Lakshmanan, K. Raum, M. Wicke, D. Mörlein, S. Brand, in *Sound Velocity and Attenuation of Porcine Loin Muscle, Backfat and Skin*. Proceedings of the International Federation for Medical and Biological Engineering, vol. 25, issue 13 (2010), pp. 96–99
M. Kompis, H. Pasterkamp, Y. Oh, Y. Motai, G.R. Wodicka, in *Spatial Representation of Thoracic Sounds*. Proceedings of the 20th Annual EMBS International Conference, vol. 20, issue 3 (1998), pp. 1661–1664
M. Kompis, H. Pasterkamp, G.R. Wodicka, Acoustic imaging of the human chest. Chest **120**, 1309–1321 (2001)
D. Leong-Kon, L.G. Durand, J. Durand, H. Lee, in *A System for Real-Time Cardiac Acoustic Mapping*. Proceedings of the 20th Annual EMBS International Conference, vol. 20, issue 1 (1998), pp. 17–20
C. Lessard, M. Jones, Effects of heart valve sounds on the frequency spectrum of respiratory sounds. Innov. Technol. Biol. Med. **9**(1), 116–123 (1988)
G. Liistro, D. Stanescu, C. Veriter, Pattern of simulated snoring is different through mouth and nose. J. Appl. Physiol. **70**(6), 2736–2741 (1991)
R. Loudon, R.L.H. Murphy, Lung sounds. Am. Rev. Respir. Dis. **130**(4), 663–673 (1984)
A.W. McCombe, V. Kwok, W.M. Hawke, An acoustic screening tool for obstructive sleep apnoea. Clin. Otolaryngol. **20**(4), 348–351 (1995)
E. Meyer, E.G. Neumann, *Physical and Technical Acoustics* (in German: Physikalische und technische Akustik) (Friedrich Vieweg & Sohn, Braunschweig, 1975)
R. Mikami, M. Murao, D.W. Cugell, J. Chretien, P. Cole, J. Meier-Sydow, R.L. Murphy, R.G. Loudon, International Symposium on lung sounds. Synopsis of proceedings. Chest **92**(2), 342–345 (1987)
M. Moerman, M. De Meyer, D. Pevernagie, Acoustic analysis of snoring: review of literature. Acta Otorhinolaryngol. Belg. **56**(2), 113–115 (2002)
A.K. Ng, T.S. Koh, E. Baey, T.H. Lee, U.R. Abeyratne, K. Puvanendran, Could formant frequencies of snore signals be an alternative means for the diagnosis of obstructive sleep apnea? Sleep Med. **9**(8), 894–898 (2008)

W.W. Nichols, M.F. O'Rourke, *McDonald's Blood Flow in Arteries: Theoretical, Experimental and Clinical Principles* (Hodder Arnold Publication, London, 2005)
H. Pasterkamp, S. Patel, G.R. Wodicka, Asymmetry of respiratory sounds and thoracic transmission. Med. Biol. Eng. Comput. **35**(2), 103–106 (1997a)
H. Pasterkamp, S.S. Kraman, G.R. Wodicka, Respiratory sounds, advances beyond the stethoscope. Am. J. Respir. Crit. Care Med. **156**(3), 974–987 (1997b)
Y. Peng, Z. Dai, H.A. Mansy, R.H. Sandler, R.A. Balk, T.J. Royston, Sound transmission in the chest under surface excitation: an experimental and computational study with diagnostic applications. Med. Biol. Eng. Comput. **52**, 695–706 (2014)
T. Penzel, G. Amend, K. Meinzer, J.H. Peter, P. Wichert, MESAM: a heart rate and snoring recorder for detection of obstructive sleep apnea. Sleep **13**(2), 175–182 (1990)
J.R. Perez-Padilla, E. Slawinski, L.M. Difrancesco, R.R. Feige, J.E. Remmers, W.A. Whitelaw, Characteristics of the snoring noise in patients with and without occlusive sleep apnea. Am. Rev. Respir. Dis. **147**(3), 635–644 (1993)
R.M. Rangayyan, *Biomedical Signal Analysis: A Case-Study Approach*. IEEE Press Series in Biomedical Engineering (Wiley Interscience, New York, 2002)
M.B. Rappaport, H.B. Sprague, Physiologic and physical laws that govern auscultation, and their clinical application. The acoustic stethoscope and the electrical amplifying stethoscope and stethograph. Am. Heart J. **21**(3), 257–318 (1941)
H. Rauscher, W. Popp, H. Zwick, Quantification of sleep disordered breathing by computerized analysis of oximetry, heart rate and snoring. Eur. Respir. J. **4**(6), 655–659 (1991)
D.A. Rice, Sound speed in pulmonary parenchyma. J. Appl. Physiol. **54**(1), 304–308 (1983)
T.D. Rossing, *Springer Handbook of Acoustics* (Springer, New York, 2007)
B. Saletu, M. Saletu-Zyhlarz, *What You Always Wanted to Know About the Sleep* (in German: Was Sie schon immer über Schlaf wissen wollten). (Ueberreuter, Vienna, 2001)
J. Schäfer, A simple procedure for quantitative and time coded detection of snoring sounds in apnea and snoring patients (in German: Ein einfaches Verfahren zur quantitativen und zeitcodierten Erfassung von Schnarchgeräuschen bei Apnoikern und Schnarchern). Laryngol. Rhinol. Otol. **67**(9), 449–452 (1988)
J. Schäfer, *Snoring, Sleep Apnea, and Upper Airways* (in German: Schnarchen, Schlafapnoe und obere Luftwege) (Georg Thieme, Stuttgart, 1996)
M. Sergi, M. Rizzi, A.L. Comi, O. Resta, P. Palma, A. De Stefano, D. Comi, Sleep apnea in moderate-severe obese patients. Sleep Breath. **3**(2), 47–52 (1999)
F. Series, I. Marc, L. Atton, Comparison of snoring measured at home and during polysomnographic studies. Chest **103**(6), 1769–1773 (1993)
S. Silbernagl, A. Despopoulos, *Pocket-Atlas of Physiology* (in German: Taschenatlas Physiologie) (Georg Thieme, Stuttgart, 2007)
F. Trendelenburg, *Introduction into Acoustics* (in German: Einführung in die Akustik) (Springer, Berlin, 1961)
I. Veit, *Technical Acoustics* (in German: Technische Akustik) (Vogel, Würzburg, 1996)
T. Verse, W. Pirsig, B. Junge-Hülsing, B. Kroker, Validation of the POLY-MESAM seven-channel ambulatory recording unit. Chest **117**(6), 1613–1618 (2000)
H.K. Walker, W.D. Hall, J.W. Hurst, *Clinical Methods: The History, Physical, and Laboratory Examinations* (Butterworth, Boston, 1990)
P.D. Welsby, J.E. Earis, Some high pitched thoughts on chest examination. Postgrad. Med. J. **77**, 617–620 (2001)
P.D. Welsby, G. Parry, D. Smith, The stethoscope: some preliminary investigations. Postgrad. Med. J. **79**, 695–698 (2003)
Wikipedia, Free encyclopedia (2010), http://en.wikipedia.org
K. Wilson, R.A. Stoohs, T.F. Mulrooney, L.J. Johnson, C. Guilleminault, Z. Huang, The snoring spectrum: acoustic assessment of snoring sound intensity in 1139 individuals undergoing polysomnography. Chest **115**, 762–770 (1999)
G.R. Wodicka, K.N. Stevens, H.L. Golub, E.G. Cravalho, D.C. Shannon, A model of acoustic transmission in the respiratory system. IEEE Trans. Biomed. Eng. **36**(9), 925–934 (1989)

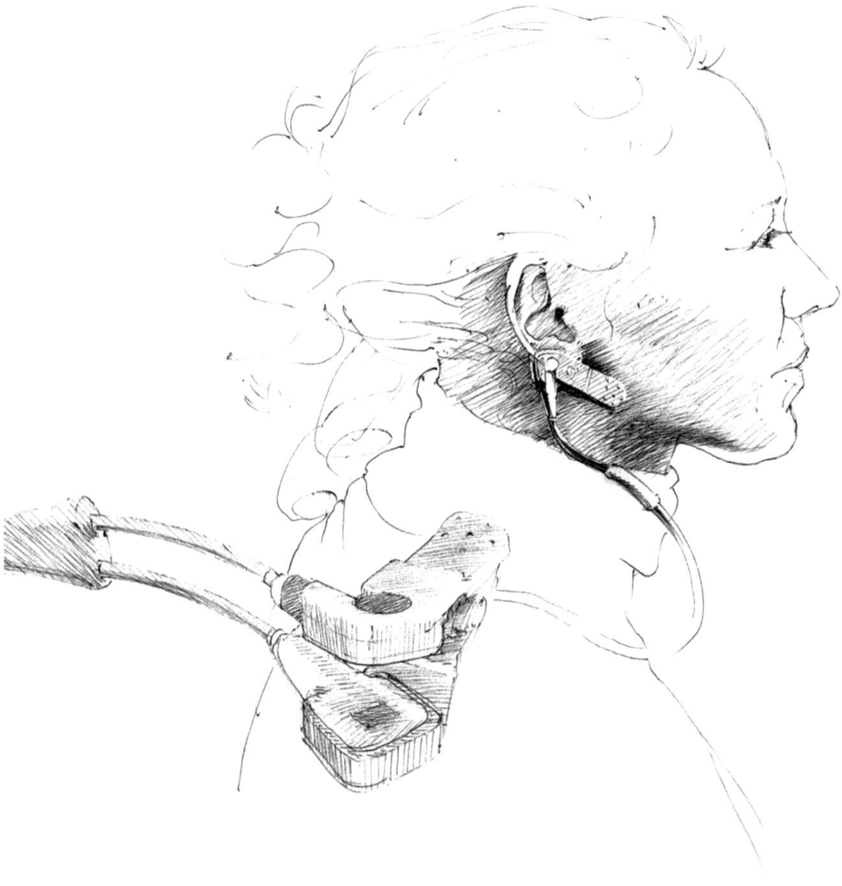

(Chamier, 2014)

Chapter 5
Sensing by Optic Biosignals

Abstract After the interface between physiologic mechanisms and the resultant biosignals has been examined (Volume I), the subsequent interface between optic biosignals and the associated sensing technology is discussed here. In the genesis of optic biosignals—induced biosignals—an artificial light is coupled into biological tissue. The resulting transmitted light intensity is strongly governed by the light absorption and scattering in tissue. The light absorption, for instance, is modulated by blood oxygenation and local pulsatile blood volume. Consequently, the transmitted light intensity reflects multiple physiologic parameters—which are vital for the assessment of cardiorespiratory pathologies and the state of health—and comprises the optic biosignal. The genesis of optic biosignals is considered from a strategic point of view. In particular, the introduced common frame of hybrid biosignals comprises both the biosignal formation path from the biosignal source at the physiological level to biosignal propagation in the body, and the biosignal sensing path from the biosignal transmission in the sensor applied on the body up to its conversion to an electric signal. Namely, the optical sensor is comprised of a light source on the skin to generate the incident light and couple it into tissue, and a distant light sink to detect the resulting transmitted light. The transilluminated region can be approximated as an arrangement of tissue layers and blood vessels. If an arterial vessel is considered with a blood pressure pulse propagating along the vessel, then there is a local pulsatile change in the arterial radius. Provided that blood in vessels absorbs the light to a larger extent compared to the tissues surrounding these vessels, it is clear that the transmitted light intensity temporarily decreases for increasing arterial radius in the transilluminated region. Thus, the propagating light is modulated by diverse physiological phenomena. A certain portion of light leaves the body and is detected by the light sink applied on the skin. The sink converts the transmitted light intensity into the electric signal. It is highly instructive from an engineering and clinical point of view how light interacts with biological tissues. Discussed phenomena teach a lot about the physics of light (as engineering sciences), and, on the other hand, biology and physiology (as live sciences). Basic and application-related issues are covered in depth. In fact, these issues should remain strong because these stand the test of time and mine knowledge of great value. Obviously, the highly interdisciplinary nature of optic

E. Kaniusas, *Biomedical Signals and Sensors II*,
Biological and Medical Physics, Biomedical Engineering,
DOI 10.1007/978-3-662-45106-9_5

biosignals and biomedical sensors is a challenge. However, it is a rewarding challenge after it has been coped with in a strategic way, as offered here. The chapter is intended to have the presence to answer intriguing "Aha!" questions.

In the genesis of *optic biosignals*, i.e., *induced biosignals* according to their classification (Sect. 1.3), an artificial light is coupled into biological tissue. The resulting *transmitted light intensity* is strongly governed by the light absorption and scattering in tissue. The induced optic biosignal—as usually registered for diagnostic and therapeutic aims—is (assumed to be) *inversely related to the transmitted light intensity* or, in other words, is (assumed to be) *proportional to the light absorption* strength.

The *light absorption*, in turn, is modulated by

- *blood oxygenation*, i.e., the saturation level of oxygenated hemoglobin in arterial blood (Sect. 3.1.4), and
- local (pulsatile) *blood volume* (Sect. 2.5.2.3).

Consequently, the transmitted light intensity reflects *multiple physiologic parameters* which are vital for the assessment of the state of health (Sect. 5.1.2.3); compare with multiparametric monitoring (Sect. 1.4).

Traditionally, optic biosignals are applied to register *blood oxygenation* and *heart rate*. Recently, medical interest has also focused on the waveform analysis of optic biosignals, facilitating, for instance, the derivation of the *respiratory rate*. The waveform of optic biosignals also indicates the state of *vascular structures* (arteries and veins) which are penetrated by the induced light.

Figure 5.1 demonstrates the basic principle of an *optical sensor* applied at a finger to deliver an induced optic biosignal. The sensor is comprised of a *light source* on the skin to generate the incident light and a distant *light sink* on the opposite site of the finger to detect the resulting transmitted light.[1] As an approximation, the transilluminated region can be given as a (series) arrangement of tissue layers and blood vessels (Sect. 5.1.2.3). If an arterial vessel is considered with a *pressure pulse propagating* along the vessel (Sect. 2.5), then there is a local pulsatile change in the arterial radius (volume) and the local ratio of blood volume to the surrounding tissue volume changes with each blood surge, compare Fig. 5.1a with Fig. 5.1b.

[1] As a curiosity concerning the *light source and light sink*, the discovery of the elementary concept of human vision is worth to be narrated shortly (Splinter 2007). It was a *great challenge* for many scientists over the centuries to find out if the human *eye emits* a "fire" seizing ambient objects with its rays and provides man with the sense of vision, or the *eye captures* light emitted by ambient sources. Aristotle (384 BC–322 BC), an ancient Greek philosopher and scientist, proposed the first hypothesis that eyes capture light. However, even many centuries later—e.g., as supported by Galen of Pergamum (around 129–200), a Greek philosopher and physician—the optic nerve was seen as a duct along which rays from the brain are emitted. It was not until the 17th century that Johannes Kepler (1571–1630), a German astronomer, provided a theory of retinal image formation.

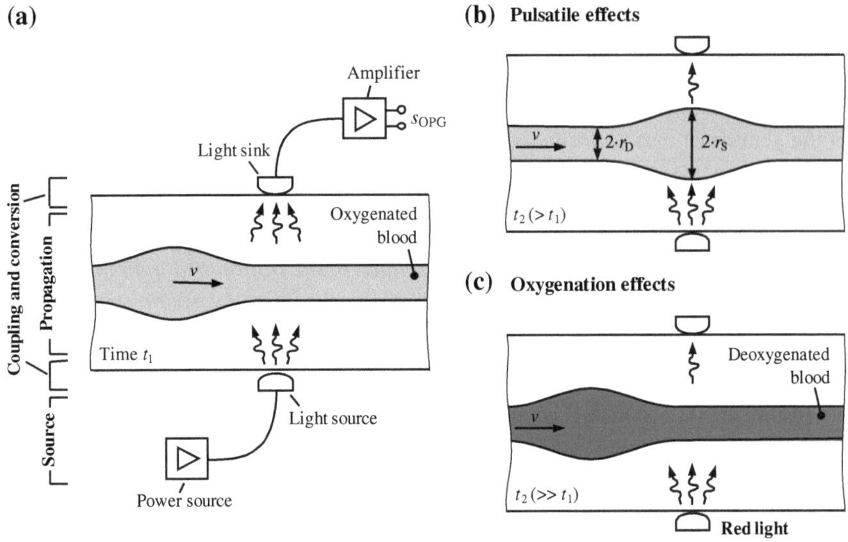

Fig. 5.1 The optical sensor applied at a finger for the registration of an induced optic biosignal optoplethysmogram s_{OPG} (Fig. 5.22). The formation and sensing paths of the optic biosignal are depicted; compare Fig. 5.2. The formation path includes an artificial light source, coupling of the incident light into the body tissue through the skin, and damping of the light during its propagation through tissues. The sensing path includes coupling and conversion of the transmitted light at the opposite skin surface. The propagation of a (pressure) pulse wave with the pulse wave velocity v is indicated for (**a**) the time instant t_1 resulting in a relatively high transmitted light intensity and (**b**) the time instant t_2 with a relatively low transmitted light intensity. Local diastolic radius r_D and systolic radius r_S are indicated (Fig. 5.14c). In addition, the transmission of red light through deoxygenated blood in (**c**) yields also a reduced transmitted light intensity in comparison with (**a**) oxygenated blood (compare Fig. 5.8)

Provided that blood in vessels absorbs the light to a larger extent compared to the tissues surrounding these vessels (Table 5.1), it is clear that the *transmitted light intensity* temporarily decreases for increasing arterial radius in the *transilluminated region*. Usually, the transmitted light intensity shows a relatively steep systolic decrease and a slow diastolic increase (Sect. 5.1.2.3). It should be repeatedly pointed out that the *local absorption of light* depends on both (global) blood oxygenation (Fig. 5.1a, c) and local blood volume (Fig. 5.1a, b). In particular, *pulsatile changes* of the local blood volume (due to the pulsatile blood pressure) are reflected by the transmitted light intensity.

The *formation of optic biosignals* up to their registration can be simplified as an *electrical circuit model*, as illustrated in Fig. 5.2 (compare Sect. 1.1). In accordance with this model and in analogy with Fig. 5.1, we start with an artificial light *source* of the incident light (represented by voltage source $\underline{U}$ in Fig. 5.2) applied on the skin

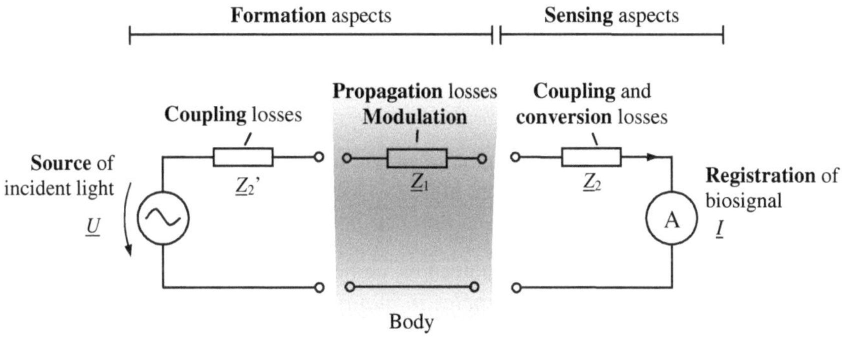

Fig. 5.2 Model of the induced optic biosignal, including its generation, coupling into the body, propagation within the body, coupling out of the body, and its registration; compare Fig. 1.3

and go over to the *coupling* of this light into the body tissue (coupling losses represented by electrical impedance $\underline{Z}_2'$ in Fig. 5.2). The *propagation* of the coupled light throughout tissue follows, in the course of which the propagating light is *modulated* by diverse physiological phenomena (modulation as electrical impedance $\underline{Z}_1$ in Fig. 5.2). As a certain portion of light leaves the body and thus is available for its detection above the skin, we continue with the light *coupling* into the light sink applied on the skin at a certain distance from the light source (coupling losses as an additive part of electrical impedance $\underline{Z}_2$ in Fig. 5.2). Lastly, the *conversion* of the transmitted light intensity into an electric signal is modelled (conversion losses as an additive part of electrical impedance $\underline{Z}_2$ in Fig. 5.2), preceding *registration* of optic biosignals (modelled as ampere meter in Fig. 5.2).

5.1 Formation Aspects

According to Figs. 5.1 and 5.2, *formation aspects* of the induced optic biosignal include

- an artificial *source* of the incident light entering the body through the skin,
- *coupling* of the incident light into the body, and
- *propagation* of light through body tissues

towards a distant light sink applied on the skin. In particular, formation aspects reveal the *modulation of the propagating light* in body tissues in synchrony with dynamic physiological phenomena. In the end, this modulation determines the physiological information hidden in optic biosignals.

5.1.1 Incident Light

5.1.1.1 Light Emission

Light[2] is emitted by an *accelerating charge* and *energy release* through downward energetic transition of electrons from higher to lower energy levels. In particular, as illustrated schematically in Fig. 5.3a, this can be attained by

- *rotation* and/or *vibration* of molecules and/or atoms (with *quantized* energy levels), or, more generally, *particles in motion*; and
- *elevation of electrons* residing in *outer electron shells* (with *quantized* energy levels) to higher energy states within the electron shell.

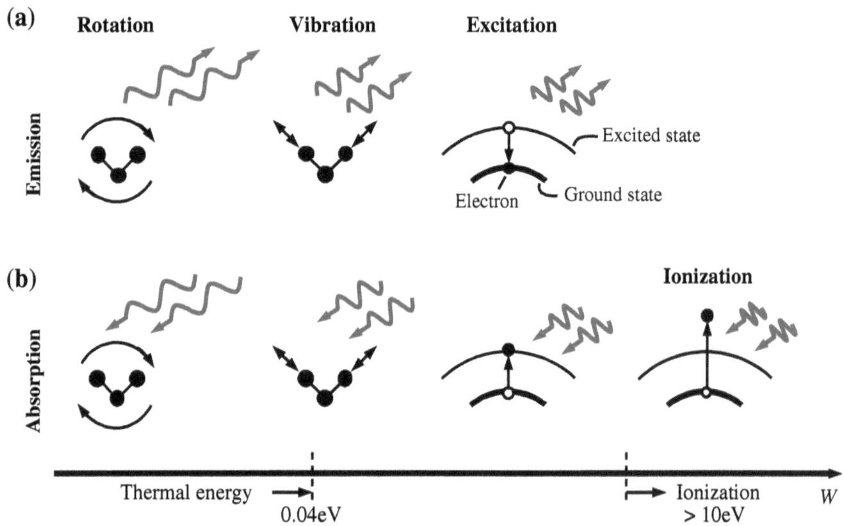

Fig. 5.3 Basic (**a**) light emission phenomena and (**b**) light absorption phenomena. These phenomena include rotation, vibration, excitation, and ionization as a function of the photon energy W (emitted or submitted energy); compare Fig. 5.4. Approximate levels of the thermal energy (at room temperatures) and ionization energy are given for comparison

---

[2] *Light* is *electromagnetic radiation* which behaves as a wave when it propagates through space (Sect. 6). In particular, electromagnetic radiation is associated with *electric and magnetic fields* which *oscillate in-phase* perpendicular to each other and perpendicular to the propagation direction of the wave; known as *transverse electromagnetic wave* (Sect. 6). If the spatial curvature of wave-forms is so small that they appear to be planar, the idealization of the wave propagation as *plane waves* is commonly used; compare Sect. 6. An important aspect of light is its *emission and absorption by accelerated charged particles* (such as electrons or polarized molecules); compare Sect. 5.1.2.

An (*uncharged*) quantum of light is known as *photon*. The photon can be seen as a minute (discrete) energy packet of light, whereas the packet's energy (i.e., the *photon energy*) depends on the frequency of the wave (5.3).

That is, the *rotational and/or vibrational motion* yield accelerated movements of electrical charges within particles in motion. These charges are linked inseparably to electric fields surrounding them; for an electric field of a single ion see Footnote 21 in Sect. 2. Thus, an *alternating electric field* is generated in time and space, which, in turn, induces an *alternating magnetic field*, in the course of which *electromagnetic waves*, i.e., the light, are created outside the particles in motion (Footnote 2); compare Fig. 5.3a. In analogy, any object with a *non-zero thermal energy*—directly proportional to the kinetic energy stored in rotational and/or vibrational motions of particles—will emit light (known as *thermal black-body radiation*[3]). In general, particles in motion prefer the lowest quantum states of energy which are available.

Concerning the *light generation* by *elevation and falling back of electrons* (Fig. 5.3), the electron structure of atom is highly relevant.[4] When an atom or a

---

[3] A (perfect) *black-body* is an object which *absorbs all radiation* incident upon it at all wavelengths and from all angles of incidence while *none of the radiation is reflected*. That is, the surface of the black-body appears colourless and black; compare Footnote 21. All radiation absorbed by the black-body is re-emitted. Interestingly, this *thermal radiation by the black-body* does not depend on the type of radiation which is incident on it but is *characteristic of this black-body* only.

In particular, the black-body (in its thermodynamic equilibrium with its environment) emits a broad continuous spectrum of electromagnetic radiation according to Planck's law (Footnote 9). The *brightness* varies with the absolute *temperature* ϑ and the *wavelength* at which it is observed. When ϑ increases, the overall *radiated energy* increases disproportionally (energy proportional to ϑ^4) while the *peak of the emission spectrum* moves to shorter wavelengths (the peak wavelength is proportional to $1/\vartheta$).

For instance, the wavelength of maximal radiation of the *sun* is about 483 nm (*visible radiation*) which corresponds to thermal black-body radiation with ϑ of 6,000 K. On the other hand, the *human skin* with ϑ of 30 °C (Fig. 3.21a) emits maximal radiation at about 9.6 μm (*infrared radiation* invisible to the human eye). In general, warm biological bodies emit infrared radiation.

[4] The *electron structure of atoms* should shortly be reviewed. All electrons in the orbit of an atom occupy the *lowest energy states* possible, i.e., from the lowest energy state (with the *strongest binding* of electrons to the positive nucleus) upwards to higher energy levels depending on the total number of available electrons. It is referred to as the *ground state* in which electrons are spatially closest to the nucleus.

In the case of *molecules* (built out of atoms) the involved molecules, atoms, nuclei, and electrons (from outer shells) *interact* with each other, so that energy states of interacting atoms are completely different (modified) in comparison to those of *isolated atoms*. Likewise, molecules and atoms in solid bodies, liquids, and dense gasses—as well as in *biological tissues*—*can not be considered as isolated*, single units in their (photon-related) *excitations* and local (phonon-related) *mechanical motions*. Additional quantized *energy levels are introduced* because of (quantum mechanically) *interacting electrons and atoms*, and, on the other hand, *vibration* of atoms of a molecule to each other and *rotation* of a molecule as a whole (Fig. 5.3).

As illustrated in Fig. 5.4, possible *energy states of a molecule*—in contrast to those of isolated atoms—are usually *composed* of

- numerous *electron states* because of interacting electrons and atoms (Fig. 5.4a). These states, in turn, are *split* into
- *vibration energy sub-states* because of atom *vibration* to each other (Fig. 5.4b). These vibration sub-states, in turn, are *split* into
- *rotation energy sub-states* because of molecule *rotation* as a whole (Fig. 5.4b).

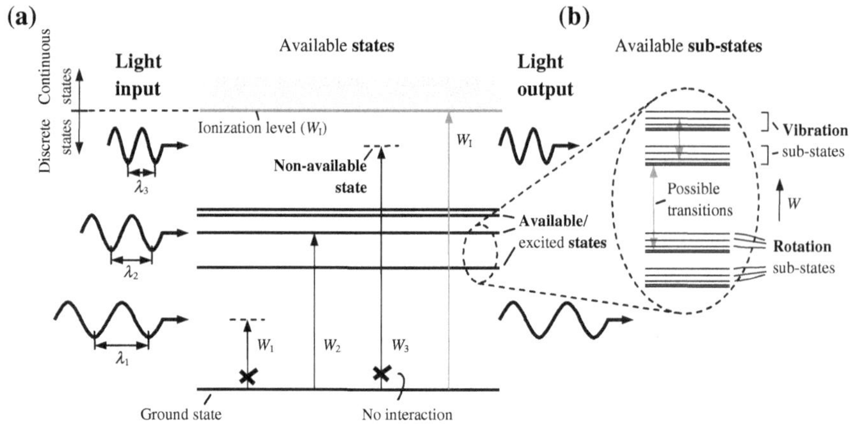

Fig. 5.4 Energy states and sub-states of a molecule (Footnote 4). (**a**) The physical principle of the light absorption with the wavelength λ_2 only ($\lambda_1 > \lambda_2 > \lambda_3$) using quantized (discrete) energy W states ($W_1 < W_2 < W_3$); compare (5.1) and (5.3). Above the ionization level W_I the possible energy states become continuous; compare Fig. 5.3b. (**b**) Rotation and vibration sub-states which compose energy states

molecule is subjected to an *external energy field* (e.g., due to electrical, optical, or chemical excitation), one or more electrons may be *excited* from their respective *ground state* to higher energy states within available electron orbits (Fig. 5.3);

(Footnote 4 continued)

Electron states occur in typical distances of about 1–10 eV. *Numerous* and closely *aligned vibration sub-states* arise in typical distances of about 0.1 eV while *rotation sub-states* arise in typical distances of about 0.001 eV (Giancoli 2006). The discrete energy levels of an isolated atom progressively *mutate* into *energy bands* with almost *continuous energy levels* when numerous atoms are brought together to form a bulk; compare Footnote 6 and Fig. 5.4. In short, *molecules* show spectra of *multiple bands* while *isolated atoms* show spectra of *multiple lines.*

In fact, the available *number of discrete electron states* within an isolated atom is much less than within a crystal or molecule including this particular atom. This number of states increases proportionally with the (usually large) number of *neighbouring atoms* which interact with this particular atom. Likewise, a single discrete energy state within an isolated atom is said to be split into numerous states when this particular atom is embedded within a crystal.

It is interesting to note that *widening of single spectral lines* also occurs, i.e., widening of individual energy states and sub-states. This is mainly due to a *limited lifetime* of excited states and sub-states, as well as due to *interactions and collisions* among involved atoms and among involved molecules. In addition, rotation and vibration sub-states are *damped* in their motions because of *interactions* among closely aligned atoms and molecules, which also contributes to the widening of individual spectral lines.

Lastly, the *Doppler effect* contributes to the *widening of spectral lines* because energy states of moving molecules *virtually change* with respect to the motionless laboratory system when these molecules move relative to the ambient electromagnetic field (or light field); see Footnote 214 in Sect. 3. That is, *moving molecules* experience higher or lower field frequencies when they move towards the light source or away from the source, respectively, and thus molecules experience the correspondingly higher or lower energy of light, in line with (5.3) (especially in gases).

provided a *match* between the *energy gap* from the ground state to a higher energy state, and, on the other hand, the submitted *external energy*. The excited electrons in higher energy states tend to return to their ground states after a particular lifetime (typically less than a microsecond); compare Fig. 5.3a. Once the excited *electrons return back*, they become *accelerated* and thus generate an alternating electric field. As already described, the alternating electric field induces an alternating magnetic field, so that *electromagnetic waves* and thus light are emitted. Likewise, the submitted energy is converted into the emitted photon energy proportional to the *energy gap* between the excited state and ground state. When the external energy is continuously provided, there is a potential for a continuous light generation (e.g., in lasers).

The *size of energy* should shortly be discussed from different perspectives and put into a relationship with the typical quantized energies of rotation, vibration, and excitation (Footnote 4). *Thermal energy* at *room temperature* is about 0.04 eV, whereas the needed energy for the excitation of atoms, i.e., *elevation of electrons* from their *ground state*, is usually > 1 eV. It indicates the stability of the atomic structure at room temperatures. On the other hand, thermal energy in *semiconductors*[5] at room temperatures is sufficient to elevate some electrons from the *valence band* into the *conduction band*,[6] i.e., is sufficient to generate electron-hole pairs. For comparison, the *binding energy* of *strong chemical bonds* (such as covalent bonds, ionic bonds, or metallic bonds) is about 1–5 eV, whereas the binding energy of *weak chemical bonds* (such as van der Waals bonds or hydrogen bonds, Footnote 18 in Sect. 2) is only about 0.04–0.3 eV (Giancoli 2006). It means that weak chemical bonds can easily (and temporarily) be broken by thermally induced collisions of molecules but not strong chemical bonds. Likewise, strong bonds hold liquids and solids together as a bulk substance at room temperatures.

---

[5] For comparison, the *valence band* and *conduction band* (Footnote 6) overlap in *conductors*. The overlap essentially yields free valence electrons which energy can easily be increased through their *acceleration* (e.g., by applied electric fields) and so electrons can easily occupy the conduction band and contribute to the electric current through conductor. The *energy gap* between the valence and conduction band is very large in *insulators* (5–10 eV) so that at ordinary room temperatures no electron (with the kinetic/thermal energy of about 0.04 eV) can reach the conduction band. For comparison, the *energy gap* is relatively narrow in *semiconductors* (about 1 eV).

[6] The *valence band* refers to the highest energy levels (states) within the electron shell of atoms, which still are completely filled with electrons. In contrast, the *conduction band* refers to available energy levels (above the valence band), in which electrons are already unbound from their individual atoms and can freely move within the atomic lattice of the material. For instance, such unbound movements contribute to the charge transport and thus to the *electric current* through the material. It should be stressed that the valence and conduction band are composed of closely aligned (quantized) energy levels, so that a single band *appears as a continuum*; see Footnote 4.

Just as an *electron* at one energy level in a particular atom may *elevate to another empty energy level* within the electron shell, so an electron can change from one energy level in a given energy band to another level in *the same energy band* or even to *another energy band*. In the latter case, the electron crosses the *energy gap* of forbidden energies provided that both bands do not overlap; compare Footnote 5.

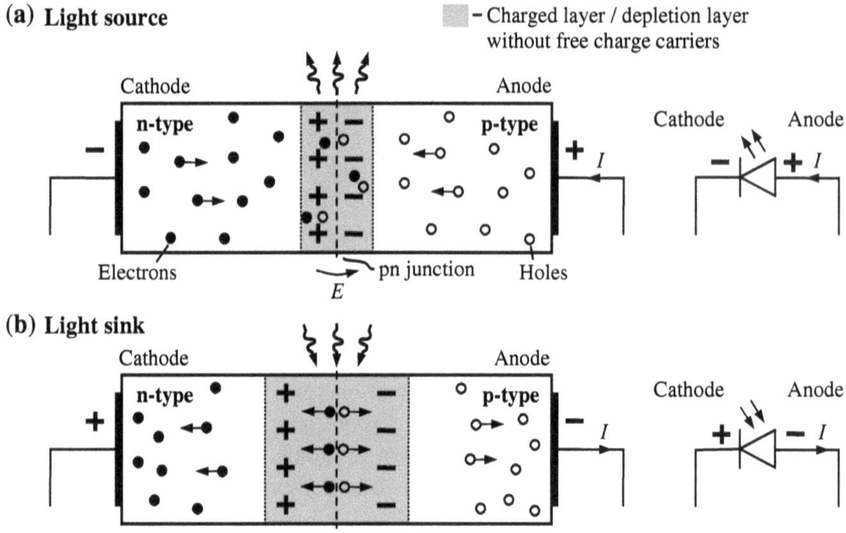

Fig. 5.5 The physical principle of (**a**) a light-emitting diode as a light source, and (**b**) a photodiode as a light sink. Here *E* denotes the electric field and *I* the electric current through the pn junction. The corresponding circuit symbols are given in the *right subfigures*

5.1.1.2 Light Source

In general, light can originate from different *light sources*. *Broadband light* sources which emit light in a relatively wide band of the electromagnetic spectrum include incandescent lamps and noble gas arc lamps. *Narrowband light* sources include lasers, fluorescent sources, and *light-emitting diodes (LEDs)*.

In fact, the *LED* comprises a *widely used light source* to induce optic biosignals (Fig. 5.1). Thus the basic principle of the LED, known as *electroluminescence*, will be described in some more detail. As illustrated in Fig. 5.5a, the LED is made of semiconductor materials that *emit light* at the *pn junction*[7] when an *electric current crosses* this junction.

[7] The boundary between the n-type and p-type semiconductor forms the *pn junction* (Fig. 5.5a).

- The *n-type semiconductor* carries an excess of valence electrons. For instance, an *impurity* atom with 5 valence electrons is embedded into the silicon lattice, so that 4 valence electrons of the impurity form 4 covalent bonds with the neighbouring silicon atoms and the *remaining valence electron* of the impurity remains weakly bound. This remaining valence electron can be easily liberated at room temperatures and is able to *carry an electric current*, like electrons in metallic conductors (Footnote 5). Likewise, the n-type semiconductor yields surplus *free negative charge carriers* (or current carriers).
- The *p-type semiconductor* lacks valence electrons. For instance, an *impurity* atom with 3 valence electrons is embedded into the silicon lattice, so that all 3 valence electrons of the impurity form 3 covalent bonds with the neighbouring silicon atoms and the *remaining unsatisfied bond* of the silicon atom creates a deficient valence electron. This unsatisfied bond

In particular, a *charge migration* (i.e., charge diffusion along the concentration gradient of charge, compare section "Transport of Substances" in Sect. 2.1.2.1) results across the *pn junction disconnected* from *external voltage*. Namely, *electrons diffuse* out of the n-type region and fill some holes in the p-type region (form negative ions). Consequently, the p-type region nearby the pn junction becomes negatively charged and the n-type region nearby the pn junction becomes positively charged because of missing electrons here (formed positive ions); compare Fig. 5.5a. An arising *electric field in the pn junction* (Fig. 5.5a)—or an intrinsic electric voltage of about 0.7 V, compare (2.5)—provides a *built-in electrical driving force* which opposes progressively the diffusional force. A *(thermal) equilibrium* of both forces occurs, preventing further diffusion of electrons across the pn junction; compare with the zero electrochemical driving force from Sect. 2.1.3.1. A *charged layer* (also known as *depletion layer*) is formed at the pn interface (with the width of about 0.1 μm), where all *free charge carriers are removed* by the local recombination of electrons and electron holes.

When the LED (or the pn junction) is *forward-biased* by an *external voltage*, i.e., the positive electrode is connected to the p-side (or *anode*) and the negative electrode to the n-side (or *cathode*), as depicted in Fig. 5.5a, the effective width of the *charged layer* decreases. Likewise, the (isolating) *depletion layer* without free charge carriers lessens in comparison to an unbiased pn junction (without any external voltage). Moreover, the *applied voltage* attracts charge carriers towards the opposite electrodes, i.e., assists

- negative electrons in crossing the charged layer towards the positive electrode, and
- positive holes in crossing the charged layer towards the negative electrode.

Likewise, the applied voltage repels electrons (holes) from the negative electrode (positive electrode). In electrical terms, the *electric current starts to flow* from the p-side to the n-side or, in analogy, from the anode to the cathode (known as operation with *forward bias*, Fig. 5.5a).

When an *electron* (coming from the n-side) meets a *hole* (from the p-side) in the pn junction, they *recombine* in this interface (i.e., the electron fills the hole). During the recombination, the electron falls into a lower energy level, i.e., from the conduction band into the valence band. The corresponding *energy difference* is released as *light photon* (Fig. 5.5a).

(Footnote 7 continued)
can attract an electron from the neighbouring (silicon) bonds at room temperatures, thus leaving a hole at the original position of the attracted electron. This hole can again attract another electron from the neighbouring (silicon) bonds to restore this unsatisfied bond. Consequently, the hole can move around the silicon crystal and is able to *carry an electric (positive) charge and electric current*. Likewise, the p-type semiconductor yields deficient negative charge or surplus *free positive charge carriers* (or current carriers), known as *electron holes*.

It should be noted that the n-type and p-type semiconductors are electrically neutral (uncharged) as a bulk substance in their ground state. In terms of energy *bands* (Footnote 6), a *hole* is a vacant (unoccupied) energy level within a band.

- The *wavelength* λ_g (and thus the colour) of the emitted *LED light* is inversely related to the width of the *energy gap* W_g between the *valence band* and *conduction band* in the *pn junction* (i.e., $\lambda_g = h \cdot v/W_g$, see 5.3); compare Footnote 5 and Sect. 5.2.1.3. The *colour* of the emitted light can be changed by varying the composition of the used semiconductor materials, i.e., by varying the width of the *energy gap*. At the same time,
- the *intensity* of the emitted *LED light* depends on the magnitude of the electric current through the pn junction (or simply through the LED, see Fig. 5.5a).

5.1.2 Transmission of Light

The *transmission of light* throughout biological tissue, after light has been coupled from an artificial light source (Sect. 5.1.1) through the skin into tissue, comprises *formation aspects* of the induced optic biosignal, according to the model of the induced optic biosignal (Fig. 5.2). As illustrated in Fig. 5.1, the optical path of light begins with the *light source*. The induced light diffuses through tissue and is subjected to changes in its intensity because of the light *absorption*, *scattering*, *diffraction*, *reflection*, and *refraction*. In fact, a large percentage of the light intensity dissipates on the way and never reaches the skin surface where a *light sink* resides (Fig. 5.1).

The *light-tissue interaction* is determined by the *qualitative type* of this interaction, the *quantitative strength and duration* of the interaction, and the *spatial localisation* (or spatial distribution) of the interaction in tissue (compare with attributes of the elementary coding of physical stimuli from Sect. 2.2.2). The interaction is usually limited to tissue areas that the coupled light managed to reach, i.e., the interaction follows the spatial pattern of the light distribution in tissue.

In fact, the *light-tissue interaction* depends strongly on both

- *light characteristics* such as the size of λ of the incident light and
- (macroscopic) *tissue characteristics* which determine the light transmission.

5.1.2.1 Propagation of Light

Light propagates in a biological medium with the *light propagation velocity* v (also known as phase velocity, a *time-space* characteristic), oscillates with the frequency f in the time domain (a *time* characteristic), and oscillates with the wavelength λ along its propagation path (a *space* characteristic) according to

$$v = \lambda \cdot f; \tag{5.1}$$

compare 4.3 and 2.21. The value of v is determined by electric and magnetic properties of the propagation medium according to

Table 5.1 Typical indices of refraction (for the wavelength of about 633 nm), absorption coefficients and reduced scattering coefficients (for about 800 nm, i.e., the isosbestic point in the oxyhemoglobin and deoxyhemoglobin spectra, see Fig. 5.8), and anisotropy coefficients (for about 630–660 nm)

Propagation medium	Index of refraction n (1) for 633 nm	Absorption coefficient μ_A (1/mm) for 800 nm	Reduced scattering coefficient μ_S' (1/mm) for 800 nm	Anisotropy coefficient g (1) for 630–660 nm
Skin* (epidermis and dermis)	1.38	1.2×10^{-2}	1.9	0.85
Skin* (subcutaneous fat)	1.45	8.2×10^{-3}	1.1	0.7
Muscle	1.37	2.8×10^{-2}	0.7	0.97
Tissue◇	1.36	10^{-2}	1	0.7–0.96
Blood	1.38	0.2□	1	0.99
Bones	1.55	2×10^{-2}	2	0.93
Water (pure)	1.33	2×10^{-3}	$< 4 \times 10^{-7}$	0▲

Rough average values are given for the different biological media such as skin (Lai 2010; Simpson 1998b; Meglinski 2002; Bashkatov 2005), muscle (Lai 2010; Simpson 1998b; Cheong 1990), (average) tissue (Tuchin 2005; Zourabian 2000; Cheong 1990), blood (Sardar 1998; Hillman 2002; Prahl 1999; Cheong 1990), bones (Genina 2008; Xu 2001; Ugryumova 2004), and (pure) water (OBOR 2003; WET 2005). The listed numbers can be compared only in a qualitative way because they have been taken from different studies which applied different measurement techniques and different model assumptions. For instance, the values of μ_A and μ_S' can be experimentally assessed by measurements of the total transmission and diffuse reflection. An additional measurement of either unscattered transmission (collimated transmission) through a thin sample or angular distribution of the scattered light permits the estimation of g (Cheong 1990)

*Skin consists of three layers—the epidermis (thickness of about 0.1 mm), dermis (about 1 mm) and subcutaneous fat tissue (about 5 mm)

◇(Average) tissue includes typically only 5 % blood as a volume fraction which corresponds to about 110 μMol/l of hemoglobin in tissue. In fact, the amount of hemoglobin determines the light absorption by tissue; compare Fig. 5.7

□Only the absorption spectrum of hemoglobin is relevant, the absorption of the blood plasma can be neglected (Cope 1991)

▲Because of the random orientation of water molecules, the light scattering by pure water itself is isotropic ($g = 0$). However, the scattering in natural water is mainly in the forward direction (with $g > 0.9$), particularly as a result of the light diffraction on particles in natural water (Davies-Colley 2003)

$$v = \frac{c}{\sqrt{\mu_r \varepsilon_r}} \cong \frac{c}{\sqrt{\varepsilon_r}} = \frac{c}{n}. \tag{5.2}$$

Here c is the speed of light in vacuum ($= 3 \cdot 10^8$ m/s), μ_r the relative magnetic permeability ($\mu_r \approx 1$ in biological media), ε_r the relative electric permittivity ($\varepsilon_r \gg 1$ in biological media, i.e., typically $v < c$), and n the index of refraction. Please note that n is a measure of how much slower a light beam propagates in a medium compared with free space, vacuum, or air ($n = 1$). In biological *tissues* that typically have high water content (of about 70 % by mass, with $n = 1.33$ from Table 5.1), the size of n is close to but somewhat higher than 1.33 depending on the density of proteins and other constituents of tissue (Furse 2009).

Table 5.1 demonstrates typical values of n for the most relevant types of *physical and biological media*. It can be observed that only *subcutaneous fat* (low water content of only about 20 %) and *bone* (low water content of about 10 %) exhibit significantly higher values of n than water. For the sake of simplicity, only approximate values of n are given in Table 5.1 for λ of about 633 nm without considering effects of varying f (i.e., without dispersion effects), temperature, and anisotropy (different media properties in different directions).

The *energy of light* (or the energy of light photons) can exist and propagate in discrete quanta[8] only, which are proportional to the size of f, to give

$$W = h \cdot f = \frac{h \cdot v}{\lambda}. \tag{5.3}$$

Here W is the energy of a photon and h the Planck's[9] constant ($= 6.6 \cdot 10^{-34}$ J · s or $4.1 \cdot 10^{-15}$ eV · s).

5.1.2.2 Effects on Light

After basic propagation phenomena of light has been discussed in Sect. 5.1.2.1, a highly instructive *interaction of light with biological tissue* should be discussed.[10] In general, the induced light is subjected to

- *volume effects* such as *absorption* which *attenuates* the light beam propagating in a *homogenous medium*; and

---

[8] The *quantised nature of light* means that increasing the intensity of light increases the *number of photons* in the light beam but not the *photon's energy*. The individual energy of photons can only be changed by varying the light frequency f (5.3).

[9] Max Karl Ernst Ludwig Planck (1858–1947) was a German physicist who originated quantum theory and proposed a revolutionary idea that energy emitted by a resonator can only take on discrete values or quanta.

[10] For the sake of completeness, it should be pointed out that the interaction of

- *uncharged radiation* such as *light* (i.e., flow of uncharged photons, Footnote 2) with a material is relatively weak in comparison with the interaction of
- *charged radiation* such as *electron beam* (i.e., flow of negatively charged electrons) with a material.

This is because most materials, especially biological samples, are composed of *polar or charged components* (e.g., water molecules are polar, compare Footnote 12 in Sect. 2) so that the charged radiation interacts more likely with such electrically non-neutral components than does the uncharged radiation (Fig. 2.4).

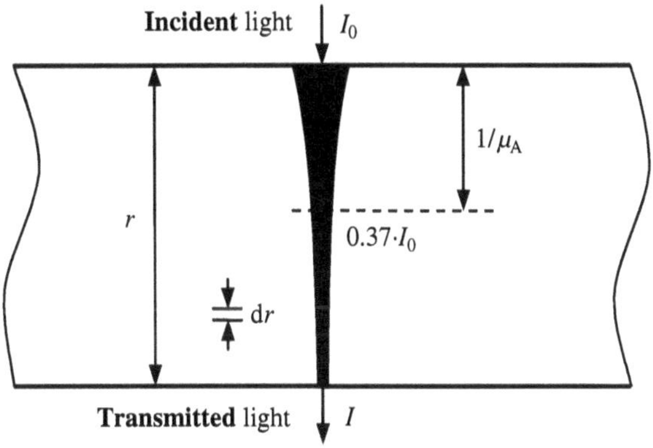

Fig. 5.6 Schematics of the light absorption with the local intensity I along the propagation distance r in a medium with the absorption coefficient μ_A; compare (5.4). The light penetration depth is denoted as $1/\mu_A$. The decreasing width of the light path indicates decreasing I due to a non-zero μ_A

- *inhomogeneity effects* such as *scattering*, *diffraction*, *reflection*, and *refraction*, which *attenuate and redirect* the light beam heading in a particular direction. The latter effects are primarily caused by a *heterogeneous medium* in the light propagation path (Fig. 5.9a).

Generally speaking, the above *effects are not fully independent* from each other. For instance, if a finite volume of tissue is exposed to an ambient light—compare Fig. 5.6—a part of the ambient *incident light* is already *reflected* back on the tissue boundary, another part of light is *absorbed* by tissue, and the rest is *transmitted* through this volume of tissue. In other words, the sum of reflected, absorbed, and transmitted portions of light should equal the ambient incident light.

Volume Effects

The *medium-related damping*[11] of light accounts for the different *volume effects* in *homogenous media*; compare section "Volume Effects" in Sect. 4.1.2.2. That is, the absorption of light quantifies the loss of light energy along the light propagation path in biological tissue.

---

[11] It should be noted that the *geometry-related damping*—as described in section "Specific Issues" in Sect. 4.1.2.1—is of minor relevance for the establishment of the induced optic biosignal (Fig. 5.1). This is due to the fact that both conditions of *far field*, i.e., the inequality $r \gg 2 \cdot \lambda$ (with r as the light propagation distance from the light source in the range of at least a few millimeters) and the light wave propagation as a *plane wave*, usually apply for light in tissue. Likewise, the decay ΔI (< 0) of the light intensity I along the finite propagation distance $\Delta r = r_2 - r_1$ (with $r_2 > r_1$, compare Fig. 4.21) can be neglected because $\Delta I \propto (1/r_2^2 - 1/r_1^2)$ and $\Delta I \rightarrow 0$ when $r \rightarrow \infty$; compare Footnote 25 in Sect. 4.

In analogy to the light emission (Sect. 5.1.1.1), the *absorption of light* (or the absorption of light photons) requires an *interaction* between light and tissue.[12] Generally, if the incident *photon energy* (Footnote 2) precisely matches the energy separating the particular quantum states of the material, i.e., matches the *energy gap* from the (initial) *ground state* to an available (final) *excited state* (with a particular lifetime), then the *light-tissue interaction* can take place (compare Footnote 24). In particular,

- *rotation* and/or *vibration* of molecules and/or atoms (Fig. 5.3b), or, more generally, transitions between rotational and/or vibrational quantum states (with discrete energy levels) can be induced, as illustrated in Fig. 5.4b. For instance, *electromagnetic waves*, i.e., the incident light, with its alternating electric field can exert a *torque* on polar molecules in tissue (molecules acting as electric dipoles) and thus can induce their rotation; compare Sect. 6. Likewise, the alternating electric field can deflect a charged atom within a molecule, after which the atom experiences a *restoring force* (approximately) proportional to its deflection width. Consequently, a vibration of this atom is initiated within the molecule. Furthermore,
- outer *electrons* within the electron shell can be *elevated* to higher energy levels by the incident light (Fig. 5.3b).

In general, the more *inert* is a transition between quantized energy levels, the lower is the needed *photon energy*. Likewise, the smaller is the *size of particles* under transition, i.e., the effective size of molecules or atoms (compare Footnote 21 in Sect. 2), the higher is their transition speed and frequency, and the higher is the needed photon energy (5.3). Consequently, at relatively low *frequency of light waves* the interaction between light and tissue is dominated by

- (relatively slow) *rotation* with the submitted (needed) energy quanta of only about 10^{-3} eV (Hoppe 1982); compare Footnote 4. With increasing frequency,
- (relatively fast) *vibration* starts to dominate with the needed energy quanta of about 0.1 eV. At relatively high frequency of light
- *electrons* start to be *elevated* to higher electronic energy levels with the needed energy quanta > 1 eV (Hoppe 1982), which corresponds to about $\lambda < 1$ μm (5.3).

The material will be *transparent* to light provided that there is *no interaction*,[13] i.e., there is no pair of available energy states such that the (incident) photon energy can

---

[12] Usually *transitions* between energy levels involve the *emission* of light photons (a transition towards a lower energy level, Sect. 5.1.1.1) or the *absorption* of light photons (a transition towards a higher energy level, Sect. 5.2.1.3); compare Figs. 5.3 and 5.5. It should be noted that the *emission wavelength* equals (approximately) the *absorption wavelength* given isolated excitable molecules and/or atoms without mutual interactions.

[13] An interesting example is given by *glass* (HyperPhysics 2012). In fact, it is *opaque* to *infrared* light (vibrational modes of the glass atoms are excited), *transparent* to *visible* light (no available energy levels above the ground state where electrons reside), and again *opaque* to *ultraviolet* light (available energy levels of electrons above the ground state); compare Fig. 5.4a.

excite the material from its *lower energy state to its higher energy state*. Figure 5.4a *demonstrates* that only the light of a particular wavelength λ_2 ($= v/f = v \cdot h/W_2$, see (5.1) and (5.3)) can interact with the available energy states of a molecule. In contrast, light beams of larger λ_1 ($> \lambda_2$ in Fig. 5.4a) or smaller λ_3 ($< \lambda_2$) transverse this molecule without any interaction, i.e., biological tissue composed of such molecules appears to be transparent for the latter beams.

Considering the *electromagnetic spectrum of light*, the following features can be observed (compare Sect. 5.2.1.4):

- *Infrared light* (with λ of about 780–10^6 nm); the light absorption in tissue is due to rotation and/or vibration of molecules and/or atoms, which typically induces *heat* (Footnote 17) in tissue.
- *Visible light* (with λ of about 400–780 nm); the light absorption in tissue is mainly due to elevation of electrons to higher energy levels because the (incident) photon energy matches spacings of available energy levels. Typically, there are many available energy states, thus the light absorption is relatively strong. The induced effects are usually limited to *thermal impact*[14] (Footnote 17) and *photochemical reactions.*[15]
- *Ultraviolet light* (with λ of about 10–400 nm); *near ultraviolet* (closest to the spectrum of visible light) is absorbed very strongly by *electron elevations*, i.e., near ultraviolet does not penetrate even a thin skin. *Far ultraviolet* can already induce *ionization* by knocking electrons out of atoms or molecules; compare Fig. 5.4a and Footnote 16. This is because the photon energy is sufficiently high to eject an electron and thus to generate a fragment with a net positive charge, such as a positive ion (compare Footnote 12 in Sect. 2).

Figure 5.4a demonstrates excitation and ionization phenomena leading to the absorption of light. If the submitted photon energy amounts to W_2, an *excitation* occurs and electrons are elevated within the electron shell of atom. However, if the submitted energy is at or above the *ionization level*[16] W_I then an excited electron is

[14] In extreme cases, excessive heat can even lead to a thermally induced denaturisation of proteins, known as *photocoagulation* (Footnote 17).

[15] *Photochemical reactions* are chemical reactions initiated in tissue by the absorption of light; compare Sect. 5.2.1.4. In consequence of the absorption, transient excited states of molecules and/or atoms can be created (including all rotation, vibration, and electron elevation), which then trigger specific chemical reactions. For instance, (visible) light with λ of less than about 600 nm is already able to resolve *strong chemical bonds* such as covalent and ionic bonds (with the binding energy 1–5 eV); compare typical binding energies from Sect. 5.1.1.1 and (5.3).

Specific chemical reactions in biological tissues can also be initiated after a special *light-absorbing substance* was injected. For instance, light with a particular λ may ultimately lead to death of unwanted or mutated cells which selectively retained the light-absorbing substance (this therapy is known as photodynamic therapy).

[16] The (lowest) *ionization energy* W_I is equal to the binding energy of an (outer) electron in an isolated atom or molecule. The level of W_I increases progressively as the atom loses (emits) electrons, i.e., electrons from orbitals closer to the nucleus experience greater forces of attraction and thus require progressively higher W_I.

no more bound to its atom of origin and is liberated from this atom (known as *photoelectric effect*). As illustrated in Fig. 5.3b, this atom then mutates to a positively charged ion. It should be stressed that *atoms and molecules* experience no *permanent changes* in their structures as long as the incident photon energy does not reach W_I, e.g., $W_I = 12.5$ eV for water molecules.

When light is absorbed within tissue, it is mostly converted into *molecular motion*, i.e., *heat*.[17] The photon energy is converted into mechanical energy, namely, into

---

[17] Considering *electrobiological interactions*, it is interesting to observe that the induced *heat* ΔW in tissue is

- *accumulated* within tissue, and, on the other hand,
- *actively transported* away from the local site of the light absorption to maintain *homeostasis* (Pfützner 2003); compare Sect. 3.1.5 and Sect. 6.

The accumulated heat increases primarily the tissue temperature ϑ by $\Delta\vartheta$. As the time t passes, the local *blood perfusion* increases and transports a part of the induced heat away from the impact site (the local region of the light absorption). Thus, the regulatory *vasodilation of vessels* in tissue (Footnote 130 in Sect. 2)—with their contact area A penetrating the tissue volume—performs important *thermal regulatory functions*, preventing overheating of the tissue and balancing its local ϑ.

The sum of the accumulated heat (proportional to $\Delta\vartheta$) and the transported heat (proportional to the product $\vartheta \cdot \Delta t$) can be expressed as

$$\Delta W = m \cdot c \cdot \Delta\vartheta + \alpha \cdot A \cdot \vartheta \cdot \Delta t$$

or, as a *differential equation*,

$$\frac{\mathrm{d}W}{\mathrm{d}t} = m \cdot c \cdot \frac{\mathrm{d}\vartheta}{\mathrm{d}t} + \alpha \cdot A \cdot \vartheta\,,$$

where m is the tissue mass, c the specific heat capacity, and α represents a measure of the blood flow velocity in (regulatory) dilated vessels. In fact, increasing both A and α facilitates the thermal regulation in tissue, whereas the resulting *time constant τ of regulatory processes* is determined by the following ratio, to give

$$\tau = \frac{m \cdot c}{\alpha \cdot A}.$$

For instance, a *step-wise supply* of the (heat-inducing) *power* $P \cdot \varepsilon(t)$ (= $\mathrm{d}W/\mathrm{d}t \cdot \varepsilon(t)$) into biological tissue with $\varepsilon(t)$ as the Heaviside step function (i.e., $\varepsilon(t) = 0$ for $t < 0$ and $\varepsilon(t) = 1$ for $t \geq 0$) yields the *exponential increase of ϑ* in tissue according to

$$\vartheta(t) = \frac{P}{\alpha \cdot A} \cdot \left(1 - e^{-t/\tau}\right) \cdot \varepsilon(t)\,;$$

compare Footnote 29 in Sect. 2 and the corresponding figure in Sect. 6. This *step response* $\vartheta(t)$ as a response of tissue (in its *zero initial state* in terms of the tissue temperature, i.e., $\vartheta(0) = 37$ °C) to a *step input* $P \cdot \varepsilon(t)$ can be easily derived from the above differential equation. The tissue is considered here as a system with a single input $P \cdot \varepsilon(t)$ and a single output $\vartheta(t)$.

Please note that the *initial rate* $\mathrm{d}\vartheta/\mathrm{d}t$ of the temperature increase in tissue is *proportional to P*; for details see Sect. 6. In particular, the resulting $\mathrm{d}\vartheta/\mathrm{d}t$ is inversely proportional to c of the exposed tissue. That is, the higher c, the lower is the resulting rate $\mathrm{d}\vartheta/\mathrm{d}t$ because more thermal energy can *rotation* and/or *vibration* of molecules and/or atoms. It should be recalled that the range of frequencies covered by (infrared) light is comparable with the *natural frequencies* at which atoms and molecules will rotate and vibrate without any incident light (i.e., without an applied field); consider similar energy levels of the infrared light (5.3), thermal energy at room temperature, and levels of energy quanta of rotation and vibration (see above). In fact, the kinetic energy of rotation and/or vibration is directly related to the *temperature of tissue*. The absorbed light, to a smaller extent, is also *converted into light* of another (larger) wavelength (Splinter 2007) if the quantum energy of the absorbed photon is higher than the average thermal energy of molecules in motion. From a quantitative point of view, if the thermal energy at room temperature is assumed to be about 0.04 eV (Sect. 5.1.1.1) then the latter condition is fulfilled for $\lambda < 30$ µm of the incident light ((5.1) and (5.3)).

While diverse absorption-related *microscopic phenomena* have been discussed on the molecular level, the *macroscopic impact* of the light absorption remains to be addressed. In fact, macroscopic effects are relevant for the establishment of the induced optic biosignal (Fig. 5.1). The *attenuation of the incident light* (modelled as a *plane wave* of infinite extent) in tissue is governed by Beer-Lambert[18] *absorption law*[19]:

(Footnote 17 continued)
be uptaken by tissue with higher c (for a given change in ϑ). For instance, this initial rate $\mathrm{d}\vartheta/\mathrm{d}t$ is less in *muscle* than in *fat* because the heat capacity c of muscle is larger than of fat (3,600 vs. 2,000 W · s/(K · kg) according to Pfützner (2003)). This rate $\mathrm{d}\vartheta/\mathrm{d}t$ decreases as t increases because the *inert thermoregulatory responses come progressively into action*, i.e., the local slope of the exponential increase decreases with increasing t.

In other words, the thermoregulatory response such as *vasodilation* causes the *temperature rise* $\mathrm{d}\vartheta/\mathrm{d}t$ to be *nonlinear*. The temperature rise is *linear* only before vasodilation increases the blood flow to a level high enough to reduce the rate of the temperature rise; this linear range may last up to a few minutes after a hyperthermia treatment has been started (Furse 2009). Given a continuous power supply into tissue and following a time interval of a few τ (e.g., after $5 \cdot \tau$), the blood flow reaches its *steady-state response*, the temperature rise converges to zero, and the temperature itself plateaus at a *steady-state level of the final temperature* $\vartheta\,(t \to \infty) = P/(\alpha \cdot A)$; see the last Equation from above.

[18] Beer-Lambert *absorption law* is named after August Beer (1825–1863), a German mathematician and chemist, and Johann Heinrich Lambert (1728–1777), see Footnote 31. The works of Beer stated that the absorptive capacity is proportional to the *concentration of the absorbing substance*. In addition, the earlier works of Lambert stated that the intensity decay of light with the *thickness of the sample* is proportional to the intensity of the incident light. Both statements yield Beer-Lambert absorption law; compare with its derivation in Footnote 19.

[19] The *rationale* of the *exponential law* in the absorption of light deserves a short notice (compare (4.7) and (5.4)). In fact, it is assumed that the light intensity drop $\mathrm{d}I$ (< 0) over the *propagation distance* $\mathrm{d}r$ is proportional to the *absorption coefficient* μ_A of light and to the light intensity I itself in the volume element with the depth $\mathrm{d}r$ (Fig. 5.6), to give:

$$\mathrm{d}I = -\mu_A \cdot I \cdot \mathrm{d}r \quad \text{or} \quad \frac{\mathrm{d}I}{I} = -\mu_A \cdot \mathrm{d}r\,.$$

Please note that the light is absorbed at a *constant rate*, i.e., $\mathrm{d}I$ depends on the total I, which, in fact, constitutes the *origin of the exponential law*. Upon integration from the *incident light intensity* I_0

$$\ln\left(\frac{I}{I_0}\right) = -\mu_A \cdot r = -\sigma \cdot \rho \cdot r \quad \text{or} \quad I = I_0 \cdot e^{-\mu_A \cdot r}. \tag{5.4}$$

Here I denotes the *transmitted light intensity* (W/m^2), I_0 the *incident light intensity*, μ_A the *absorption coefficient* (1/m) of light in tissue, and r the path length of light through tissue. In general, the absorption law assumes a homogenous propagation medium,[20] absent inhomogeneity effects (such as scattering and reflection, section "Inhomogeneity Effects" in Sect. 5.1.2.2), and a collimated incident light.

It can be observed that μ_A $(= \sigma \cdot \rho)$ is proportional to both the *absorption cross section* σ (m^2) and the number density ρ (1/m^3) of *chromophores*.[21] The area σ represents an effective area (target area) that a single chromophore presents to an incident light photon. In fact, the size of σ changes effectively with λ, whereas ρ can be assumed to be constant; compare μ_A variations as a function of λ in Fig. 5.8.

Figure 5.6 demonstrates schematically the light absorption by a bulk tissue. According to (5.4), the incident light intensity decays by about 63 % after the path $r = 1/\mu_A$; compare with a similar *exponential decay* from Fig. 4.22. It is important to note that the ratio $1/\mu_A$ can also be interpreted as an *average free path length* of light or a light photon in tissue before the photon encounters the next absorption event. Likewise, the coefficient μ_A is reciprocal of the light *penetration depth* and indicates the frequency (or number) of absorption events per unit length travelled.

(Footnote 19 continued)
to the (decreased) *transmitted light intensity* I after the distance r (on the left side of the above equation) and a simultaneous integration from the initial depth $r = 0$ to the final depth r (on the right side of the above equation), the *differential equation* from above yields the exponential law of the light absorption from (5.4).

[20] If *more than one absorber* with μ_A^1 and μ_A^2 is present, the *effective product* $\mu_A \cdot r$ (5.4) is the *sum* of the corresponding products of each absorber, i.e., $\mu_A \cdot r = \mu_A^1 \cdot r + \mu_A^2 \cdot r$. In analogy, *two absorbers connected in series*—such as a *blood vessel* with μ_A^1 embedded within (otherwise bloodless) *tissue* with μ_A^2 ($\ll \mu_A^1$, Table 5.1) in accordance with Fig. 5.1a—yield the effective product

$$\mu_A \cdot r = \mu_A^1 \cdot r_1 + \mu_A^2 \cdot r_2.$$

Here r_1 is the path length of light through the *blood volume* in the vessel and r_2 the path length through the *tissue volume* surrounding the vessel. In the case of the embedded blood vessel, it can be approximated $\mu_A \cdot r \approx \mu_A^1 \cdot r_1$, which means that *blood* in tissue or pulsatile extensions of this blood vessel determines the *total light absorption* of perfused tissue. Likewise, the product $\mu_A \cdot r$ (or $\mu_A^1 \cdot r_1$) increases temporarily from (local) diastole to (local) systole because the systolic r_1 is larger than diastolic r_1 (section "Cardiac Activity" in Sect. 5.1.2.3 and Fig. 5.14c).

[21] The *chromophore* is a chemical group, e.g., a group of atoms or molecules, that causes the chromophore's coloured appearance. The chromophore *absorbs* light at certain wavelengths which depend on available energy states of this chemical group. Namely, if the energy possessed by photons falls within the energy difference between two available energy states (*energy gap*) the photons are absorbed (compare Footnote 4 and Fig. 5.3). The remaining wavelengths are *transmitted* or *reflected*, which gives rise to a specific *observed color* (Footnotes 3 and 23).

The expression $e^{-\mu_A \cdot r}$ in (5.4) shows the *probability of photon survival* after the path length r.

Just as the structure of biological media is highly *heterogeneous* (Sect. 2), so different physical and biological media in the body all have their *unique* characteristics of the *light absorption*. Table 5.1 lists some typical values of μ_A for λ of about 800 nm. It is obvious that *blood*—with *hemoglobin as the most important chromophore* in tissue (Footnote 21)—is the *strongest absorber*; namely, μ_A of blood is at least 10 times larger than μ_A of any other biological medium. The light absorption of the (average) *tissue* is about 20 times less than that of blood because the typical *volume fraction* of blood in the (average) tissue amounts to only 5 % (or 2–5 % according to Jacques (1998)). Likewise, the fraction of *blood determines the light absorption in tissue*.

In particular, the light absorption of the skin *epidermis* (the outermost *skin* layer) is usually dominated by *melanin absorption* proportional to the *volume fraction* of melanosomes in the epidermis; melanosomes are cell organelles containing melanin (compare Sect. 2.1.1). The volume fraction is in the range of several percent, e.g., about 10 % in a moderately pigmented adult skin (Jacques 1998). Melanin exhibits *increasing absorption with decreasing* λ. It implies that melanin (and thus the epidermis) plays a significant role in the absorption of the incident *visible light* and has only little impact on the absorption of *infrared light*.

The absorption of the skin *dermis* (the skin layer below the epidermis) depends on a dominant *hemoglobin absorption* because the dermis—in contrast to the epidermis—is perfused with blood. The average *volume fraction of blood* here is only about 0.2 %. However, the cutaneous blood is unevenly distributed within the dermis yielding local dermal regions (layers) with the volume fraction of blood up to 5 %. That is, the local perfusion in the dermis can be as high as the perfusion in the (average) tissue. In general, *melanin and hemoglobin* are *principle tissue chromophores*. From a practical point of view, the intensity of the incident light should be sufficiently high for the *light to penetrate the skin* without significant losses if the (pulsatile) absorption of tissue under the skin has to be registered.

Furthermore, Table 5.1 reveals that the *subcutaneous fat* exhibits μ_A which is only about one third of μ_A of *muscle* (with a relatively high *fraction of blood*). It means that the incident light probes the adipose tissue to a greater depth than muscle; compare Fig. 5.24 with the relevant discussion on the *light probing in a heterogeneous tissue* (section "Probing Depth" in Sect. 5.2.1.1). In particular, this observation has a practical relevance if the optical sensor is applied on the (chest) skin below which a significant amount of the adipose tissue exists; compare sections "Respiratory Activity" in Sect. 5.1.2.3 and 5.2.1.1.

However, the coefficient μ_A *varies strongly over* λ and the tissue type, as depicted in Fig. 5.7. That is, a specific biological medium absorbs light strongly at specific values of λ and not at others. The shown absorption spectra, especially those of oxygenated and deoxygenated blood, prove that *red light* and *near-infrared light* (upwards from about 600 nm up to about 1,300 nm) are less susceptible to absorption by biological tissues than *blue and green light*; this is also known as *optical window* into tissue (Wang 2012). The lower boundary of this

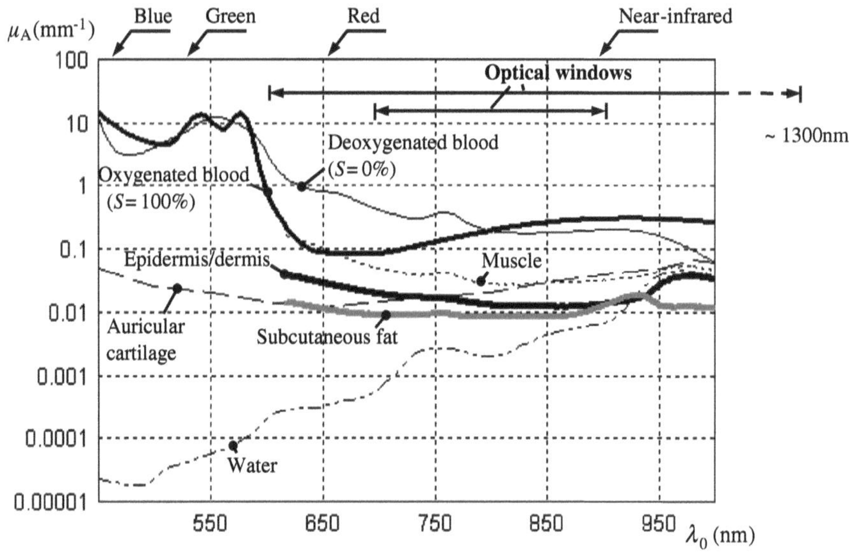

Fig. 5.7 Light absorption coefficient μ_A of different biological media as a function of the light wavelength λ_0 in free space. Data on the absorption by oxyhemoglobin and deoxyhemoglobin in blood was taken from BORL (2012), Cope (1991) with an assumed (realistic) hemoglobin molar concentration of 2.3 mMol/l (or 150 g/l) in blood; compare Fig. 5.8. The corresponding hemoglobin oxygen saturation S is indicated. Data on water was taken from Segelstein (1981), auricular cartilage (Youn 2000), epidermis/dermis, subcutaneous fat, and muscle (BORL 2012)

optical window is determined by relatively strong absorption of green and blue light by perfused tissue and pigmentation (Wukitsch 1988). The upper range of this window is mainly provided by (on average) increasing and *dominating light absorption by water* (abundant in biological tissues) with increasing λ in the *infrared range*. An even *better optical window* is given in the spectral range of 700–900 nm (Fig. 5.7), which avoids a local maximum in the water absorption at about 970 nm. Consequently, the *applied wavelengths*[22] in practice reside

[22] In fact, the particular choice of the applied wavelengths depends on a particular goal of the biomedical optics. If the light has to *permeate a heterogeneous biological tissue* then the aforementioned *optical window* for *red and near-infrared* light should be used, where the total light absorption is relatively low; compare Fig. 5.7 with the indicated colours of light.

On the other hand, if some components of this heterogeneous tissue should be contrasted with other components, e.g., blood vessels should be contrasted with the surrounding tissue (i.e., with a sort of the (average) tissue, almost bloodless tissue, see Table 5.1), for the unaided human eye then an *optical contrast* between vessels and the surrounding tissue should be aimed at Kaniusas (2011). Thus, spectral regions of light should be found where the light *absorption and scattering* by blood vessels differ as strongly as possible from the light absorption and scattering by the surrounding tissue.

Such *optical contrast*, for instance, is provided by *blue and green* light, so that the corresponding μ_A of blood dominates strongly with respect to μ_A of the different tissues (almost bloodless tissues); compare Fig. 5.7. In contrast, *red* light is absorbed by blood to a lesser degree; i.e., red light

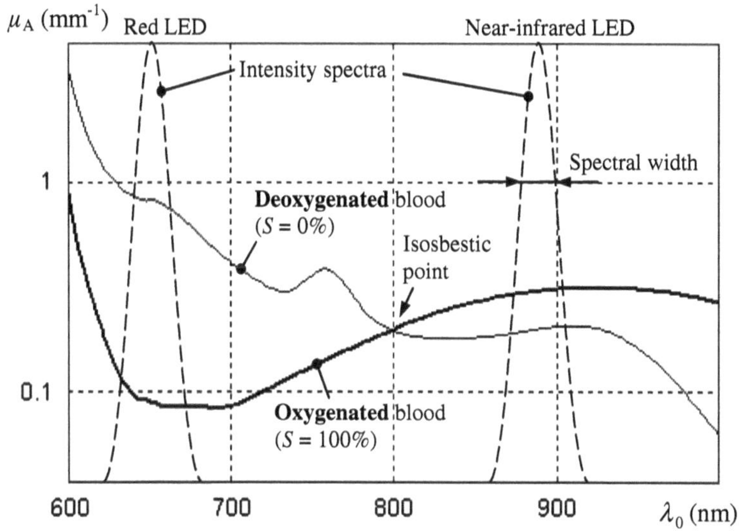

Fig. 5.8 Light absorption coefficient μ_A of oxyhemoglobin (oxygenated blood) and deoxyhemoglobin (deoxygenated blood) as a function of the light wavelength λ_0 in free space and hemoglobin oxygen saturation S; compare Fig. 5.7. The hemoglobin molar concentration of 2.3 mMol/l (or 150 g/l) is assumed in blood. In addition, schematic intensity spectra of light-emitting diodes (LEDs) for red and near-infrared light are superimposed for oximetry-related comparison purposes. Data on μ_A extracted from BORL (2012), Cope (1991)

somewhere in the range from 600 up to 1,000 nm—as indicated by emission spectra of LEDs in Fig. 5.8—so that the applied light can penetrate tissues to maximum depths of a few centimetres (section "Blood Oxygenation" in Sect. 5.1.2.3).

It can be observed that *oxygenated blood and deoxygenated blood*, namely, *oxyhemoglobin* and *deoxyhemoglobin*, dominate the light absorption in the shown spectral range of 450–1,000 nm (Fig. 5.7). It confirms the dominance of μ_A of blood in Table 5.1. In particular, deoxyhemoglobin dominates below 800 nm, whereas oxyhemoglobin dominates above 800 nm; compare Fig. 5.8. The blood dominance with respect to surrounding (almost bloodless) tissues tends to be more prominent for *blue and green* light than *red and near-infrared* light. To be precise, μ_A for red light has already fallen to less than one tenth of μ_A for blue or green light in blood. In analogy with Table 5.1 and Fig. 5.7 reveals that the light absorption by the *subcutaneous fat* is less than one third of the absorption by *muscle*.

(Footnote 22 continued)
penetrates both *bloodless* and *blood-rich* tissues with little losses and produces practically no optical contrast between blood vessels and the surrounding tissue. Considering the light *scattering*, it should be noted that the scattering strength does not vary as strongly as the absorption strength (i.e., the size of μ_A) over λ. Therefore, the scattering phenomena can hardly be applied for the aforementioned contrast enhancement by optimizing the colour of the incident light.

Figure 5.8 demonstrates an excerpt of Fig. 5.7 in which μ_A of hemoglobin is stated against λ considering two *different oxygenation levels* of hemoglobin (Sect. 3.1.4). At *red* wavelengths at about 660 nm, μ_A of oxyhemoglobin (related to the *oxygen saturation* $S = 100\ \%$ in blood) is significantly less than μ_A of deoxyhemoglobin ($S = 0\ \%$), whereas at *near-infrared* wavelengths at about 890 nm, to a lesser degree, the reverse is true. The resulting intersection point at which μ_A of oxyhemoglobin and μ_A of deoxyhemoglobin are equal is close to 800 nm and is known as *isosbestic point*. In other words, oxyhemoglobin does not absorb much red light; however, when *S drops*, increasing amount of red light is absorbed by *blood* which becomes *progressively darker*[23] (i.e., less and less red light is reflected back to an observer); this effect is used to measure *S* (section "Blood Oxygenation" in Sect. 5.1.2.3).

Please note that the ratio of μ_A to μ_A at the *isosbestic point* (Fig. 5.8) remains sensitive to the level of *S* while being less sensitive to other factors (like tissue blood volume, see section "Specific Issues" in Sect. 5.1.2.3). In analogy, the reversal of μ_A with respect to the *isosbestic point* is the key issue within the scope of spectrometry (and *oximetry*, sections "Blood Oxygenation" in Sect. 5.1.2.3 and 5.2). For instance, this reversal allows an experimental *estimation of S* independent of the total density of hemoglobin in tissue if the incident light of (at least) two different wavelengths is used; compare Figs. 5.1a, c and 5.8.

Inhomogeneity Effects

In addition to the light absorption in terms of the medium-related damping in *homogenous media* (section "Volume Effects" in Sect. 5.1.2.2), the *heterogeneous structure* of the biological tissue impacts strongly the *attenuation of light* propagating in a specific direction, e.g., towards the skin surface. In general, the light beam interacts with *microscopic structures* such as *molecules, cell organelles, membranes, and cells* (Fig. 2.2), and, on the other hand, with *macroscopic structures* such as *fibers* and *layered tissues*. A highly heterogeneous macroscopic structure of the light propagation pathway is demonstrated in Fig. 5.9a, in which the light beam crosses skin, arteries, veins, and the surrounding tissues. This *light-tissue interaction* is rather complex, in the course of which the light beam may experience *spatial redirection* and *accumulated attenuation* while heading towards an optical sensor (on the skin).

In particular, the following *effects govern the propagation of light* in tissue towards the skin surface and the coupling of light into the air above the skin (or into the light sink on the skin, Fig. 5.1):

[23] When *hemoglobin* in blood is oxygenated (*S* rises), blood absorbs *red light* to a lesser degree than when oxygen is depleted in blood (*S* drops); see Fig. 5.8. Therefore, a relatively large amount of red light is *reflected* back to the observer when oxyhemoglobin dominates, i.e., *arterial oxygenated blood* looks *reddish*. On the contrary, dominating deoxyhemoglobin in *venous deoxygenated blood* lets venous blood to appear *bluish*.

- scattering,
- diffraction,
- reflection, and
- refraction.

Scattering

The (elastic[24]) *scattering*[25] of light in tissue is due to the chaotic *variation* in the index n of *refraction* (compare (5.2)) at a microscopic and macroscopic scale,

---

[24] In fact, besides the *absorption* of light (section "Volume Effects" in Sect. 5.1.2.2), the *interaction of the uncharged radiation* (Footnote 10) with a material includes Krieger (2004)

- *elastic scattering* (see Footnote 25) and
- *inelastic scattering*.

In particular, the *absorption of the uncharged radiation* can be based on

- *complete absorption of a photon* (section "Volume Effects" in Sect. 5.1.2.2 and Fig. 5.3b) by either
  - *motions* of molecules and/or atoms, i.e., rotation and/or vibration of molecules and/or atoms, or
  - *excitation* of atoms, i.e., elevation of an electron within the electron shell of atom, or even
  - *ionization* of atoms, i.e., removal of an electron out of atom. In addition,
- *partial absorption of a photon* can also take place, known as *inelastic scattering*. In particular,
  - considering *Raman effect*, the incident photon excites *rotational and vibrational motions* of polarizable molecules. It yields scattered photons with less energy (and larger λ, (5.1) and (5.3)) by the amount of the transition energy (or the energy gap, section "Volume Effects" in Sect. 5.1.2.2).
  - In the case of *Compton effect*, the incident photon *knocks an electron* out of an atom (usually a weakly bound electron from an outer shell). Another photon is simultaneously emitted (or scattered) with less energy by the amount of the bound electron energy (i.e., of the electron removed out of the atom).
  - *(Quasi) elastic scattering* may also take place, in which the spectrum of the scattered light is *Doppler-broadened* in comparison with the spectrum of the incident light. This is because *translational and rotational motions* of (optically anisotropic) molecules shift the frequency of the scattered light depending on their motion velocity; compare Footnote 214 in Sect. 3. For instance, the effective λ of the scattered light is shortened when the scatter (or the molecule) moves towards the light sink. Correspondingly, the frequency of the scattered light increases while the light propagation velocity remains constant; compare (5.1).

However, in the case of *near-infrared* and *visible light*, the *ionization* of atoms and the *inelastic scattering* are almost irrelevant because this light has a relatively low quantum energy ($\ll$ 1 keV, (5.3)).

[25] *Elastic scattering* of the light wave occurs when *charged particles* in a medium (e.g., electrons in molecules) are forced to *oscillatory motions* by the *electric field* of the incident light wave or, in other words, by photons hitting the electron shell and inducing forced oscillations of electrons in it. These accelerated motions start to *emit light* of the same frequency as the incident wave; compare Footnote 2. The respective (re)radiated patterns from (numerous) oscillating particles *superimpose* and yield a "*single*" *source* of the scattered light. Likewise, the *photon energy* of the scattered photon is not changed in the course of the *elastic scattering*.

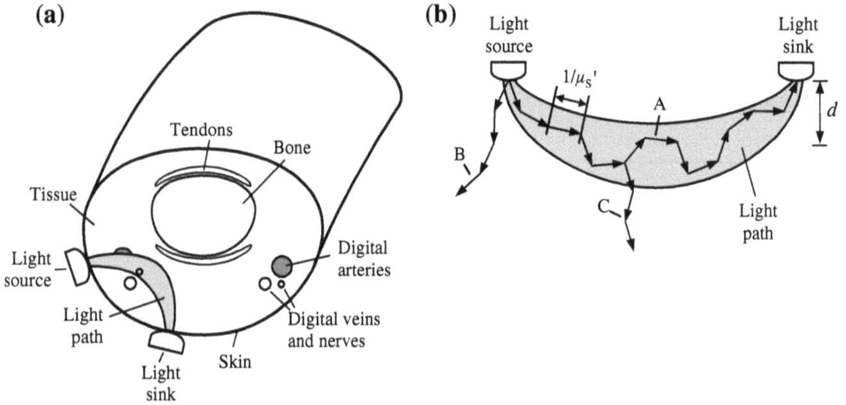

Fig. 5.9 Propagation of light in the reflectance mode (Fig. 5.22). (**a**) Cross section of a finger with the propagation of the incident light through its heterogeneous structures. (**b**) Possible random pathways of photons within tissue, whereas the banana-shaped region (*grey background*) denotes the region of photons which most likely will be detected by the light sink. The mean probing depth is denoted as *d*; compare Fig. 5.20

which results in a dispersion of light in all directions. The mismatch in n is due to (compositional) *tissue components* such as (collagen) fibers, cell membranes (the lipid-water interface, Fig. 2.3), extracellular/intracellular fluid, and cell organelles (Fig. 2.2). In addition, *fluctuations in density* and *thermal motions* of molecules contribute to the scattering. In fact, tissue components appear to merge into a continuous structure with a spatial variation in n.

As illustrated in Fig. 5.10a, the paths taken by light photons which transverse tissue are no longer direct in comparison with a direct path in Fig. 5.6. *Multiple scattering* and *multiple paths* in tissue yield the *diffuse propagation* of photons. A *forward and focused beam* of light is *attenuated* due to its widening and back-scattering. In addition, the effect of scattering substantially increases the effective *path length* travelled by photons within tissue and therefore substantially *increases the probability of their absorption* (compare (5.4)).

Photons are strongly scattered by those structures whose size matches the photon wavelength λ. As an approximation, photons are said to be *scattered by particles* of different sizes, i.e., scattered by an aggregation of material that constitutes a (homogenous) region with n_1 that differs from n_2 of the particle's surrounding.

(Footnote 25 continued)

In particular, the *incident electric field* of light forces *electric dipoles* in the *dielectric medium* (induced and permanent dipoles, see Sect. 6) to align and alternate in synchrony with this incident field. In the *conducting medium*, its *free charge carriers* (such as free electrons or ions) oscillate back and forth at the same frequency as the incident field. Likewise, *oscillating charged particles* act as *small antennas*, reradiating the incident wave that becomes the *scattered wave*.

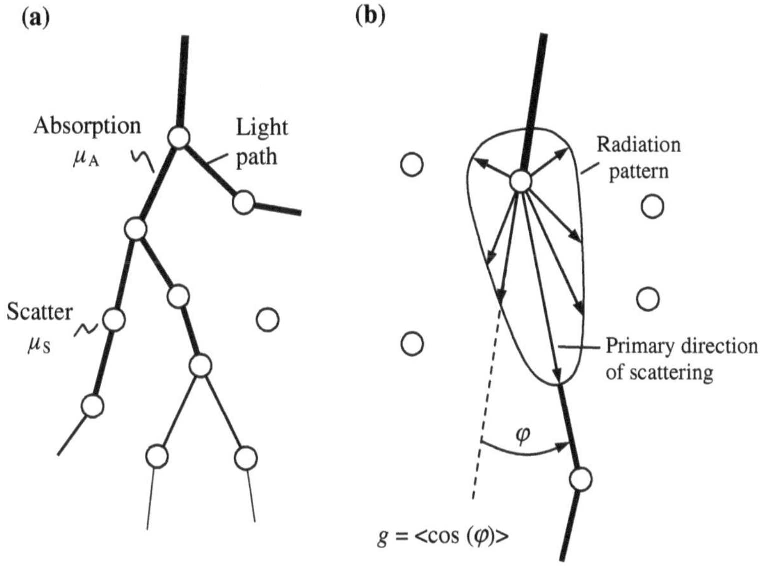

Fig. 5.10 (**a**) Schematics of multiple scattering and absorption of light. The decreasing thickness of the light path indicates decreasing intensity of the transmitted light due to a non-zero absorption coefficient μ_A and a non-zero scattering coefficient μ_S; compare Fig. 5.6. (**b**) Definition of the scattering anisotropy coefficient g as the average of cosine of the deflection angle φ; compare Fig. 5.11b

The pattern of the scattered wave[26] and the direction in which the scattered photons travel depends on the variation in n, the size and shape of the scattering particle, and the size of λ. The discussed *optical window* (section "Volume Effects" in Sect. 5.1.2.2) in the range of 600–1,000 nm (Fig. 5.8) is of high importance concerning the practical choice of λ. In contrast, when the structure size gets much larger than λ, *geometrical optics* becomes applicable and the reflection laws apply (see below).

In fact, two basic types of the scattering arise

- *Rayleigh*[27] scattering, i.e., redirection of the rays of light for particles *smaller than* λ of the incident light, and

---

26 Since there are *many scattering particles* (e.g., in the highly *heterogeneous* biological tissue), the *individual scattered waves* from each particle combine to form the *entire scattered wave* at the *observation point*. Provided that the scattering particles are randomly located in tissue, the individual scattered waves have *random phases* at the observation point; i.e., the distance light travelled from each particle to the observation point is random. It is instructive that the *power density*—proportional to the square of the resulting *electric field* magnitude, see Sect. 6—of the entire scattered wave at the observation point is equal to the sum of the power densities scattered from each particle. It is known as *incoherent light scattering*.

27 John William Strutt Rayleigh (1842–1919) was an English physicist who made significant contributions in the fields of optics and acoustics. He explained the concept of wavelength-dependent light scattering as well as the sunlight interaction with the upper atmosphere yielding blue coloured sky (Footnote 29).

- *Mie*[28] scattering, i.e., redirection of light for particles *similar to or larger than* λ of the incident light.

Rayleigh scattering refers to the scattering on the *microscopic structure* of tissue components which size is less than approximately 1/10 of λ. The scattering structures comprise mainly the *lipid-water interface* (a strong mismatch in n, see Table 5.1) in membranes of the cell (Fig. 2.3) and membranes of the cell's organelles such as mitochondria (Fig. 2.2) with the structure size of about 9 nm (Jacques 2002). Likewise, the *water-protein periodicity* of *collagen fibers*—with the corresponding mismatch in n of 1.33 and 1.5 (Jacques 1998)—with the displacement size of about 70 nm contributes to Rayleigh scattering; this size is still much smaller than the applied λ (>600 nm in the optical window, Fig. 5.8). As illustrated in Fig. 5.11a, small structures scatter in all directions with a constant intensity, known as *isotropic scattering*. A *small particle* experiences an almost homogenous electric field over its entire size ($\ll \lambda$) so that this particle emits (scatters) light as a point source (Footnote 25). There is a complete symmetry in the forward and backward scattering. Likewise, light beams *scattered from the different parts* of a single small particle (Fig. 5.11a) *interfere constructively* with each other at the *observation point*, so that each part scatters *in-phase* with the scattering of adjacent parts (Footnote 25). The *scattering intensity scales inversely with* λ as $(d/\lambda)^4$, where d is the effective size of the scattering particle. That is, Rayleigh scattering is strong at short wavelengths.[29] On the other hand, *large particles* are more *efficient scatterers* than small particles.

Mie scattering refers to the scattering on the *microscopic structure* of tissue, which size is similar to or larger than λ. Such relatively large structures comprise protein aggregates, nuclei of cells, lysosomes, mitochondria (about 1 μm in size) or even *collagen fiber bundles* (about 3 μm in diameter) of the skin (dermis). Mie theory predicts that scattering by micrometer-sized particles is strongest when the particle radius and λ are of the same order of magnitude (Tuchin 2005). As illustrated in Fig. 5.11b, relatively large structures scatter mainly in the forward direction within (relatively) small angles of the incident beam axis, known as *anisotropic scattering*. In contrast to Rayleigh scattering, a *large particle* experiences a heterogeneous electric field over its entire size ($\geq \lambda$) so that the different

[28] Gustav Adolf Feodor Wilhelm Ludwig Mie (1868–1957) was a German physicist who described the scattering of light by particles whose size is comparable to the wavelength of light.

[29] The *blue colour of the sky* at daylight is caused by the *wavelength dependent Rayleigh scattering* off molecules of the air. The blue end of the visible light is scattered more effectively in the direction down to the earth, i.e., in the direction which is almost perpendicular to the direction of the sun's incident light. In other words, the *isotropic Rayleigh scattering* (Fig. 5.11a) dominates towards an observer on the earth, who looks from *overhead*, not directly into the sun. In contrast, the *anisotropic Mie scattering* (Fig. 5.11b) is mainly *directed forward* in the direction of the sun's incident light and thus does not reach this observer. Consequently, the blue of the sky is less saturated when the observer looks closer to the sun because the forward directed *Mie scattering* then starts to dominate. Mie scattering is less strongly wavelength dependent than Rayleigh scattering and thus scatters in the whole range of the visible light, not only at its blue end. Likewise, relatively large *water droplets* in a *cloud* scatter in the whole visible range, which causes the *cloud to appear white*; compare Footnote 30.

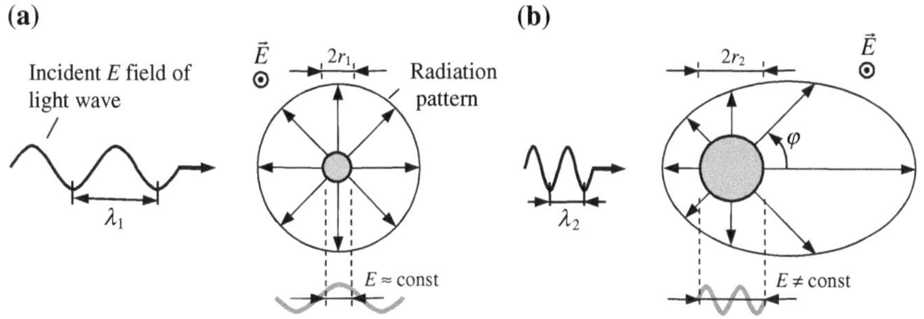

Fig. 5.11 Radiation patterns of light scattering in the plane perpendicular to the incident electric E field. (**a**) Isotropic Rayleigh scattering of light with the wavelength λ_1 on a relatively small particle with the diameter $2r_1$ ($\lambda_1 > 2r_1$). E field is approximately constant throughout the entire particle because $\lambda_1 > 2r_1$. The resulting radiation pattern (polar plot) shows the angular dependence of the scattering strength for a spherical particle. (**b**) Anisotropic Mie scattering on a relatively large spherical particle ($\lambda_2 \leq 2r_2$) with strong dependence of the scattering strength on the deflection angle φ and with the varying E throughout this particle; compare Fig. 5.10

parts of this particle are excited with the different phases and thus emit (scatter) light with the different phases (Footnote 25). There is a *constructive and destructive interference* of scattered waves, i.e., scattered from the different parts of this single particle, at the *observation point* in certain directions, depending upon geometrical conditions. Likewise, contributions to the total scattered wave—emanating from different parts of the large particle towards the observation point—are travelling over distances that are a significant fraction of λ (or even a few λ) apart, which yields *significant phase variations* of these contributions and thus the constructive and destructive interference. Consequently, the aforementioned *angular dependence* of the intensity of the scattered light results, whereas local peaks and valleys at various angles may even occur. Figure 5.11b illustrates the varying intensity of the scattered light over the scattering angle φ. Mie scattering is less strongly wavelength dependent than Rayleigh scattering; e.g., white glare (but not blue, Footnote 29) can be observed around the sun if a lot of relatively large particles are present in the air. In fact, the *scattering intensity scales inversely with* λ as $(d/\lambda)^c$ where d is again the effective particle size and c is (only) somewhere in the range 0.5–1.5 (Jacques 2003). Therefore, the power of Mie scattering increases when d gets larger[30]; however, this power increase is less than that in the case of Rayleigh scattering (due to $c < 4$, see above).

[30] The observation that the *scattering efficiency* increases with the *particle size* in both *Rayleigh* and *Mie scattering* can be observed in *cataract*, i.e., cloudy vision and glare, especially in aged persons. Here special proteins aggregate in the *lens* (damage of lens's proteins due to their unfolding and denaturation, see Footnote 18 in Sect. 2), which results in the local density variations leading to local variations in the *index of refraction*. Thus *scattering centers* are built in the lens, which scatter light out of its normal path to the retina. Blurred and dimmed images result. In addition, halos around point-like light sources can be also seen as a consequence of the forward scattering; compare Footnote 29.

Figure 5.12 demonstrates *radiation patterns* of the scattered light after a *thin collimated beam* of the incident light has passed a thin layer of the *skin*. The radiation patterns are given as the normalised intensity of the transmitted light plotted as a function of the *deflection angle* φ, the angle between the deflected light and the normal to the skin surface. As expected, the maximum intensity arises in the direction of the normal, i.e., for $\varphi = 0$, and then decreases with increasing magnitude $|\varphi|$. Interestingly, these patterns differ significantly and conclusively for two λ of light and two thicknesses of the penetrated skin layer, as revealed in Fig. 5.12:

- For a relatively *thin skin layer* (Fig. 5.12a), it can be observed that the scattering is mainly forward directed (*weak scattering*) for a relatively long λ (of *infrared light* at 1,700 nm in Fig. 5.12a). The corresponding radiation pattern decreases steeply with increasing $|\varphi|$. In contrast, the scattering is much broader (*strong scattering*) for a relatively short λ (of *visible light* at 600 nm in Fig. 5.12a). The corresponding radiation pattern follows *Lambert's*[31] *cosine law* (Hardy 1956), i.e., the radiation decreases in proportion to the cosine of φ and thus disappears for $\varphi = \pm 90°$. This observed wavelength dependence is in line with the *inverse scaling of the scattering with* λ, as discussed above; i.e., infrared light is less scattered than visible light. As illustrated in Fig. 5.12a, the scattered light with the angle φ' (> 0 from the normal) yields a higher intensity I' for $\lambda \approx 600$ nm (strong scattering) than for 1,700 nm (weak scattering); i.e., the inequality $I'_{600} > I'_{1700}$ applies. Consequently, the transmitted intensity in the forward direction ($\varphi' = 0$ in Fig. 5.12a) is larger for infrared light than visible light if incident intensities of both lights are identical.
- For a relatively *thick skin layer* (Fig. 5.12b), it can be observed that the radiation pattern of the scattering is almost independent of λ and follows *Lambert's cosine law*. This is because the number of *multiple scattering* events—the light within the skin layer is subjected to—increases with increasing thickness of the layer. The number of scattering particles along the pathway of light throughout the layer increases with increasing thickness (Fig. 5.10). Consequently, even *infrared light* mutates from forward directed (Fig. 5.12a) into broadly spread (Fig. 5.12b) after having passed a relatively thick skin layer. The authors in (Hardy 1956) demonstrate that the scattering is already maximal—and thus its radiation pattern is almost independent of λ—for skin samples with the thickness > 1 mm for all λ from the range of 550–2,400 nm.

Different biological media involve particles of *different sizes* and *different types* that requires consideration of both *Rayleigh scattering* and *Mie scattering*. Their particular predominance depends on the discussed relation between the structure size and λ (Fig. 5.11). In general, the *angle of scattering* is a function of the size and

[31] Johann Heinrich Lambert (1728–1777) was a Swiss German physicist, mathematician, astronomer, and philosopher who, among others, pioneered work in photometry and proved the irrationality of π (= 3.14159 …, i.e., the ratio of a circle's circumference to its diameter); compare Footnote 18.

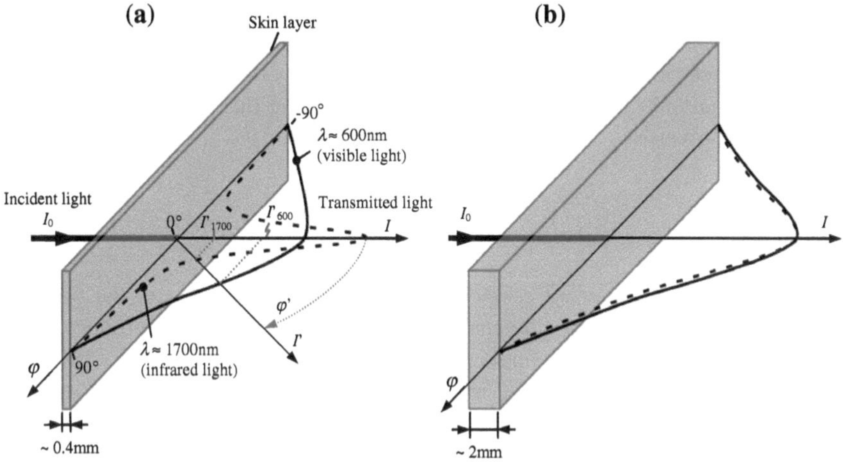

Fig. 5.12 Scattering of a single collimated light beam with the incident light intensity I_0 after its passage through a single skin layer. Radiation patterns for two wavelengths λ (600 and 1,700 nm) show the behaviour of the transmitted light intensity (in relative units) as a function of the varying deflection angle φ (varying from $-90°$ to $90°$ from the normal to the skin surface). Data on radiation patterns was taken from Hardy (1956). (**a**) Relatively thin skin layer. (**b**) Relatively thick skin layer

shape of the particle, the size of the incident λ, and the angle at which the incident light reaches the particle surface.

It is important to stress that the *lipid-water interface* plays a major role in the scattering processes. Rayleigh scattering is mostly induced by the *cellular membranes* while Mie scattering by the membrane folds and *membranous structures* such as mitochondria as a whole. The light scattering—as discussed above—increases with decreasing λ, i.e., it is much stronger in the blue region than the red, as much as the *wavelength dependence* of the light absorption which is stronger in the blue region than the red (Fig. 5.7). However, the scattering strength—as a rough approximation—does not vary as strongly as μ_A over λ.

Furthermore, soft tissues with higher or lower *lipid content* show increased or decreased scattering, respectively (Jacques 2002). In the relevant spectral region from the visible to near-infrared light both Rayleigh and Mie scattering affect the light scattering in the *skin*.[32] The impact of the *heterogeneous tissue* composition,

[32] In the *skin*, Rayleigh scattering from small-scale structures dominates below 650 nm while Mie scattering from large-scale structures dominates above 650 nm (Jacques 1998). The *epidermis* of the skin is mainly composed of *keratin fibers* which behave like collagen fibers in the dermis (Jacques 1998). The thinness of the *epidermis* reduces its *scattering relevance* for visible and near-infrared applications involving the light diffusion. Thus the scattering in the *dermis* alone can be used to describe the skin scattering in general terms. In the *dermis*, *Rayleigh scattering* is induced by the microstructure within *collagen fibers* and other small cellular structures, whereas *Mie scattering* is induced by the relatively large and elongated *collagen fibers* (section "Inhomogeneity Effects" in Sect. 5.1.2.2).

especially of subcutaneous adipose and muscular layers under the skin (Fig. 5.24), on the light scattering and light probing in tissue is discussed in section "Probing Depth" in Sect. 5.2.1.1 in detail.

The *attenuation of light* in its *propagation direction* due to a *single scattering event* can be given as

$$\ln\left(\frac{I}{I_0}\right) = -\mu_S \cdot r \quad \text{or} \quad I = I_0 \cdot e^{-\mu_S \cdot r}, \tag{5.5}$$

whereas μ_S is the *scattering coefficient*. In analogy with (5.4) and the provided interpretation of μ_A, the coefficient μ_S can be interpreted as the product of the scattering cross section of a single scattering particle and the number density of scattering particles. Likewise, μ_S is reciprocal of the *average free path* of light without its scattering (i.e., before a photon encounters the next scattering event) and indicates the frequency (or number) of scattering events of a photon per unit length travelled. The expression $e^{-\mu_S \cdot r}$ shows the *probability of the photon transmission* without redirection by scattering after the path length *r*.

The level of *anisotropy* (Fig. 5.11b) is accounted by the *scattering anisotropy coefficient g*. It describes the *asymmetry of the scattered light* after a single scattering event. The *level of g* is a measure of the amount of the deflected light which is aligned in the direction of the incident light. If the angle φ is the angle between the incident photon and deflected photon—as illustrated in Fig. 5.10b—then the component of the deflected photon (or the new trajectory) which is aligned in the incident (forward) direction is given by the projection $\cos(\varphi)$. For a light beam, i.e., a number of photons, there is an *average deflection angle* and the average of $\cos(\varphi)$ is defined as *g* for two dimensional light propagation (Elwell 1999; Cheong 1990; Jacques 2003):

$$g = \langle \cos(\varphi) \rangle = \int_{2\pi} p(\varphi) \cdot \cos(\varphi)\, d\varphi. \tag{5.6}$$

Here $p(\varphi)$ is the probability density function which describes the angular distribution of the deflected light for a single scattering event (or the relative probability of the scattering with the angle φ); see Fig. 5.10b. The value of *g* approaching 1, 0, −1 describes highly *forward scattering*, *isotropic scattering*, and highly *backward scattering*, respectively. The size of *g* varies typically with λ.

Figure 5.10b demonstrates that the *scattering occurs in all directions* including the backward direction. Thus the values of φ encompass the full range from 0 to 2π for two dimensional light propagation; compare (5.6). The length of arrows (outwards from the scattering point in Fig. 5.10b) represents the radiation strength in this particular direction. Likewise, the most likely scattering angle φ corresponds to the longest arrow and thus to the *primary direction of the deflected light*. Table 5.1 summarizes *g for different biological tissues* showing there is fairly *strong anisotropy*. It can be observed that the scattering in tissues occurs principally in the *forward direction* with $g > 0.7$.

The *reduced scattering coefficient* μ_S' is another lumped parameter according to

$$\mu_S' = \mu_S \cdot (1 - g) \tag{5.7}$$

which describes the random diffusion (random walk) of photons involved within numerous *isotropic scattering* events (Fig. 5.11a). The *average free path* of light or the step size of the resulting random walk is $1/\mu_S'$ where each step involves a scattering event; compare Fig. 5.9b. Likewise, incident photons are converted into scattered photons within the depth $1/\mu_S'$ below the incident surface. This parameter μ_S' can be regarded as an effective *isotropic scattering coefficient* (i.e., $\mu_S' = \mu_S$ for the isotropic scattering with $g = 0$, while $\mu_S' = 0$ for the forward scattering with $g = 1$) that represents the cumulative effect of several (anisotropic) forward-scattering events.[33] In fact, the level of μ_S and g of biological tissues depend on the spatial *mismatch* of n. For fully matched n of scatters and background medium (the scattering vanishes), the coefficient μ_S' goes to zero while g approaches unity (forward propagation only). In addition, it should be noted that a simultaneous occurrence (or a relative contribution) of *Rayleigh scattering* with the relevant $\mu_S'(\mathrm{R})$ and *Mie scattering* with $\mu_S'(\mathrm{M})$ contributes additively to the total coefficient $\mu_S' = \mu_S'(\mathrm{R}) + \mu_S'(\mathrm{M})$.

As already discussed within the scope of *Rayleigh and Mie scattering*, the coefficients μ_S *and* μ_S' *increase with decreasing* λ (Simpson 1998a; Franceschini 1999a; Zourabian 2000). The spectral dependence of μ_S' was reported to be less pronounced in the peripheral adipose tissue than deeper in the muscular tissue (Franceschini 1999a).

Table 5.1 summarizes μ_S' *for different biological tissues*. It can be observed that upper *skin* layers—especially, the *dermis* of the skin (Footnote 32)—and bones comprise the strongest scatters. This table also confirms the aforementioned strong scattering by *lipids*, i.e., the listed value of μ_S' for the subcutaneous fat is higher than that for the *muscle*. In general, *water* is a weak scatter; i.e., μ_S' of pure water is very low so that the light scattering has a small influence compared to the absorption by water.

The light *scattering* at the applied red and infrared wavelengths (Fig. 5.8) is by far the most *dominant tissue-photon interaction* compared to the light absorption (Elwell 1999). The *collimated light*[34] *becomes diffuse light*—or the incident light loses all of its original directionality—already at low *penetration depths* of usually

[33] Provided that the light *scattering dominates over its absorption* and the light propagates mainly in the *forward direction*, i.e., $\mu_S \gg \mu_A$, $\mu_S' \gg \mu_A$, and g is close to 1, a description of photon movements with a single (relatively) *large step* $1/\mu_S'$ is equivalent to another description with many (relatively) *small steps* $1/\mu_S$ (Jacques 2002); please note that $\mu_S' < \mu_S$ if $g > 0$ (5.7). A large step with $1/\mu_S'$ involves only *isotropic deflection* while each small step involves *anisotropic deflection*. This situation with the scattering as the dominant *tissue-photon interaction*, known as *diffusion regime*, usually applies to biological tissues exposed to visible light and near-infrared light (Table 5.1). In addition, the diffusion regime *justifies the use of* μ_S' (5.7).

[34] For the *transmission* of a *collimated light* through a *thin tissue layer* the *total absorption coefficient* $\mu_T = \mu_A + \mu_S$ can also be defined. Here the total absorption of the incident light intensity is governed by (5.4) with μ_T instead of μ_A. For *thicker tissue layers*, the collimated light

a *few millimeters*; compare Fig. 5.23. Despite the predominantly forward scattering, this is because μ_S (within the typical range 10–100 mm^{-1}) is approximately two orders of magnitude greater than μ_A; compare data on μ_S' from Table 5.1 with the conversion from (5.7). In other words, only some of the incident light is absorbed in tissue while a large portion of light is scattered. Likewise, a single photon encounters (on average)

- *many scattering events before* this photon is absorbed by
- *a single absorption event*;

compare Footnote 33. It allows the light to *penetrate tissue to greater depths* in the range of a few centimetres and thus to extract physiological information from (relatively) deep tissues (Sect. 5.2.2).

Provided that the *light scattering dominates over the absorption* (Footnote 33), the anisotropy of biological tissue is not very strong, and the distance between the light source and light sink (compare Fig. 5.9a) is relatively large, i.e., about many diffusion lengths of light (compare (5.9)), then the *photon diffusion theory* can be applied (Jacques 2003; Schmitt 1991; Cheong 1990). This theory accounts simultaneously for the absorption and scattering effects in tissue. The attenuation of the incident light can be given as

$$\frac{I}{I_0} = c \cdot \frac{e^{-\alpha \cdot r}}{4\pi \cdot D \cdot r} \tag{5.8}$$

where c is a constant, D the (isotropic) diffusion length defined as

$$D = \frac{1}{3 \cdot \left(\mu_A + \mu_S'\right)}, \tag{5.9}$$

and α the *total light attenuation coefficient* which is defined as

$$\alpha = \sqrt{3\mu_A \cdot \left(\mu_A + \mu_S'\right)}\ . \tag{5.10}$$

In analogy with (5.4) and Fig. 5.6, the coefficient α is reciprocal of the light *penetration depth* (see Fig. 5.23). Please note that α increases with increasing μ_S' so that, as expected, α increases from the forward scattering (g = 1 and μ_S' = 0 yield a relatively low α, see (5.7)), to the isotropic scattering (g = 0) and then to the backward scattering ($g = -1$) with the least light arriving at the light sink.

As already noticed, *multiple scattering* substantially increases the *effective path length* of light (Fig. 5.10a), particularly because the scattering dominates in tissue. Consequently, the effective *attenuation of light increases*, which can be accounted by the so-called *differential pathlength factor*, a scaling factor of the geometrical

(Footnote 34 continued)
transmittance can also be described by an exponential law; however, *multiple scattering* should be accounted for (Tuchin 2005).

propagation distance r. This factor accounts for the additional distance travelled by photons. Likewise, the distance r in (5.4) can be substituted by the *true optical distance* ($> r$)—i.e., the product of the differential pathlength factor (> 1) and r—which constitutes the *modified Beer-Lambert absorption law* (Hillman 2002; Elwell 1999). To give an example, the differential pathlength factor in the adult head has a value of approximately six (Elwell 1999). This high value shows the importance of the scattering in human tissues.

Diffraction

In analogy with the diffraction of body sounds (section "Inhomogeneity Effects" in Sect. 4.1.2.2) and electric biosignals (Sect. 6), the *diffraction of light* waves manifests as the apparent *bending of waves* around small particles, i.e., around an aggregation of material with n different from the particle's surrounding (compare Fig. 4.25). Here *small particles* mean that their size is small compared to the size of λ or on the order of λ:

- If the size of λ *is large* in relation to a particle (or the size of structures in tissue), the larger part of the incident wave will *readily diffract* around the particle and remains unaffected. Likewise, for longer λ the particle behaves as a point source of diffracted waves, whereas the particle's shape is of little importance. In analogy, the light can also spread out beyond *small openings* (Footnote 41 in Sect. 4), i.e., small compared to λ.
- If the size of λ *is in the same order* of magnitude as the size of the particle, the *diffraction* (bending of waves towards the particle past this particle, compare Fig. 4.25b) becomes less pronounced. In addition, the *interference* usually arises among diffracted waves creating (alternating) regions of greater light intensity (due to constructive interference) and lesser light intensity (destructive interference); compare Footnote 41 in Sect. 4.
- For an even *smaller* λ below the size of the particle, most of the incident light is reflected back according to the following *reflection laws* (5.11) and a *light shadow* is formed behind this particle; compare section "Inhomogeneity Effects" in Sect. 4.1.2.2. The *diffraction is small* and can be neglected in the behaviour of the propagating wave; i.e., ray tracing and geometrical optics can be used. For instance, the shadow of a large object ($\gg \lambda$) shows small fringes near its edge because of the residual diffraction.

Reflection

The *reflected light* bounces back into the incident medium while the *refracted light* enters another medium behind the boundary of the two media, see Fig. 5.13. The *reflection laws* state—in analogy with the reflection of electric biosignals (Sect. 6) and body sounds (section "Inhomogeneity effects" in Sect. 4.1.2.2)—that the *incident angle* to the normal (to the reflective surface at the point of the incidence) equals the *reflection angle* to the normal, whereas all three the incident wave, the

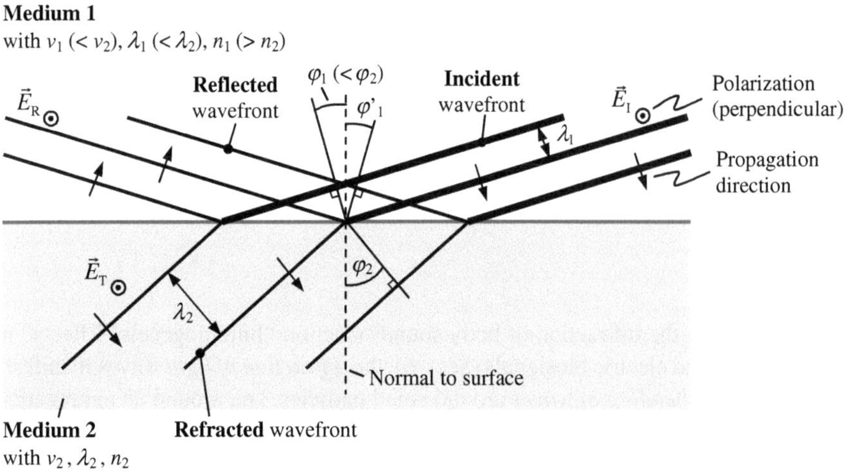

Fig. 5.13 Reflection and refraction of light on the boundary of two different media, shown in the plane of incidence. Here v is the corresponding light propagation velocity, λ the light wavelength, n the index of refraction, φ_1 the incident angle (in the slow medium 1), φ'_1 the reflection angle ($= \varphi_1$, in the medium 1), φ_2 the refraction angle (in the fast medium 2), and $\vec{E}$ the electric field with the perpendicular polarization to the plane of incidence (Footnote 35). The *line thickness* of wavefronts indicates roughly the light intensity

reflected wave, and the normal lie in the same plane. Figure 5.13 demonstrates the reflection of light on the *boundary* between two media, whereas the equality of the incident angle φ_1 and the reflection angle φ'_1 can be observed. It should be reiterated that the *reflection laws apply* only if *body dimensions are much larger than* λ of the incident light.

The amount of the reflected wave (as related to the incident wave) from a medium with n_1 *normally incident* on another medium with n_2 ($\varphi_1 = \varphi'_1 = 0$ in Fig. 5.13) is determined by the *optic reflection factor*[35] Γ_O, given by

$$\Gamma_O = \frac{E_R}{E_I} = \frac{n_1 - n_2}{n_2 + n_1}. \tag{5.11}$$

Here E_R and E_I are the respective amplitudes (= peak values) of the reflected and incident (oscillating) electric field of light; compare Footnote 2. A negative value of Γ_O would indicate that the reflected light wave, i.e., the electric field of the wave,

[35] The *reflection factor* depends on both the *angle of incidence* and the *plane of light polarization*, i.e., the direction of the electric field $\vec{E}$ can be *parallel or perpendicular* to the plane of incidence. In particular, (5.11) describes the reflectivity at *normal incidence* ($\varphi_1 = 0$ in Fig. 5.13) and the *perpendicular polarization* (see $\vec{E}$ in Fig. 5.13) only.

experiences a *phase reversal*[36] in relation to the incident light wave (the reflected wave is 180° out-of-phase from the incident wave). Equation (5.11) yields $\Gamma_O > 0$ for the depicted case in Fig. 5.13 with $n_1 > n_2$; thus there is *no phase shift* between E_R and E_I fields in Fig. 5.13.

For instance, Fig. 5.23 demonstrates *the course of* $|\Gamma_O|$ as a function of λ for optical radiation incident on the surface of the *human skin*. It can be observed that the skin reflects predominantly *visible radiation* with increasing λ, whereas *infrared radiation* is reflected up to about 1,400 nm; compare section "Organs at Risk—Eye and Skin" in Sect. 5.2.1.4.

Interestingly, as the *thickness of the skin* (from skin samples in vitro) increases, the amount of the *reflected light*—from the light incident on the skin surface—tends to increase, e.g., the factor $|\Gamma_O|$ increases by about 0.1 or 10 % with increasing skin thickness from 0.4 to 0.8 mm (i.e., doubling in thickness) at $\lambda = 1$ µm (Hardy 1956). It indicates that the light is not only reflected from the outer skin surface but also *back-scattered from inner skin layers* residing under the outer skin surface. In addition, Fig. 5.23 reveals that the reflection minima and maxima almost coincide with minima and maxima of the *penetration depth* $1/\alpha$, respectively. Likewise, the following relationship is even more informative; the *reflection minima and maxima* coincide with *maxima and minima of the total attenuation* α. This relationship also indicates that

- *strong reflections* are due to strong back scattering from *non-absorbing inner skin layers* or, conversely,
- *weak reflections* are due to weak back scattering from *absorbing inner skin layers*.

Refraction

The *refracted light* experiences a redirection (or *bending*) owing to a change in n and thus in v (5.2). In analogy with the refraction of body sounds (section "Inhomogeneity Effects" in Sect. 4.1.2.2), when light passes from a *slow medium* into a *fast medium*—as shown in Fig. 5.13—the refracted light is bent away from the normal. It can be observed that the distance between neighbouring *wavefronts* increases in the fast medium which implies that the *wavelength* λ increases (i.e., $n_2 < n_1$ and $\lambda_2 > \lambda_1$; see (5.1)); compare Sect. 6. It is important to note that the *frequency* f of light does not change when light enters another (linear) medium.

---

[36] When light *reflects* from a medium of *higher n* ($n_2 > n_1$ and $\Gamma_O < 0$, (5.11)), the reflected light experiences a *phase shift of 180°* on the boundary (light with the perpendicular polarization only, Footnote 35). Otherwise, there is *no phase shift* when light reflects from a medium of *lower n* ($n_2 < n_1$ and $\Gamma_O > 0$), as also illustrated in Fig. 5.13. Such phase shifts play an important role in the light interference which, for instance, manifests colourfully in thin soap films.

Similar behaviour was already observed in Fig. 4.26 where reflected *body sounds* also experience a phase shift of 180°. This is because inner body sounds bounce back into tissue from the tissue-air boundary, whereas the air shows a lower sound propagation velocity ($v_A < v_T$ in Fig. 4.26) and thus a higher index of "acoustical" refraction (compare (5.2)).

Provided that the dimensions of the refracting surface are larger than λ of light, the laws of the *optic refraction* apply. In fact, the amount of bending is subject of *Snell's refraction law* (Footnote 44 in Sect. 4)

$$\frac{n_1}{n_2} = \frac{\sin(\varphi_2)}{\sin(\varphi_1)}, \tag{5.12}$$

where φ_2 is the *refraction angle*. The incident wave, the refracted wave, and the normal lie in the same plane. Equation (5.12) yields the inequality $\varphi_2 > \varphi_1$ for the depicted case in Fig. 5.13 with $n_1 > n_2$, which implies that the refracted wave (in the medium with n_2) is bent away from the normal.

5.1.2.3 Light Modulation by Physiological Phenomena

As already illustrated in Fig. 5.1, the propagating light in *tissue* undergoes specific absorption and scattering in the course of which the transmitted light intensity is reduced but also obtains an *alternating component*. That is, *light* is *dynamically modulated* by diverse physiological phenomena in perfused tissue (Fig. 5.2) because these phenomena modulate *optical properties of tissue*.

Numerous *light absorbers* are present in tissue, which can be roughly subdivided into

- *pulsatile absorbers* such as *pulsatile arterial blood*; and
- *non-pulsatile absorbers* such as *non-pulsatile arterial blood*, *capillary blood*, *venous blood*, and bloodless *tissue*; compare Fig. 5.9a.

Figure 5.14 demonstrates all these absorbers comprising the so-called *compartmental model of living tissue* (Zourabian 2000; Schmitt 1991). The absorbing components in this model (Fig. 5.14b) are schematically related to a typical time course of the *absorption strength* (Fig. 5.14a), i.e., the time course of the *product* $\mu_A \cdot r$ (5.4) over a few *cardiac cycles*. As a practical approximation, the term $\mu_A \cdot r$ is inversely related to the *transmitted light intensity I*.

With each blood surge (or after each *ventricular contraction*) a pressure pulse reaches the transilluminated region where the local light absorption increases due to increased *local blood volume* in the (arterial) capillary beds[37] (Fig. 5.1a, b).

[37] In the cutaneous tissue, slight *dilation and contraction of arterioles and capillaries* during each pressure pulse contribute to the local changes in the light absorption. Consequently, a high *density of arterioles and capillaries* is required near the surface of the skin—as found e.g., in the fingertip—to attain large *pulsatile changes in the absorption*.

It seams that *arterio-venous-anastomoses* (i.e., shunts between arterioles and venules, see Fig. 3.22b), *venules*, and *veins* contribute to the pulsatile volume in the *cutaneous circulation* (Kim 1986). The *shunting* of a (high pressure) arterial pulse via *open arterio-venous-anastomoses* generates a *pulsatile volume* in the (low pressure) local venous side, where changes in the vascular venous volume are much greater than those in the arterial bed. Here, it should be recalled that *venous compliance* is greater than *arterial compliance* (Sect. 2.5.1). Thus the same variation in the

As a result, the corresponding transmitted intensity I decreases temporarily (section "Cardiac Activity" in Sect. 5.1.2.3). An oscillation of I with the pulsatile frequency f_C occurs, which corresponds to the *pulsatile alternating component* I_{AC} of I (Fig. 5.14a) and to the *pulsatile arterial blood* in terms of the *compartmental model* (Fig. 5.14b). Thus the component I_{AC} is inversely related to the *pulsatile part of the total absorbance*, whereas large I_{AC} indicates strong arterial pulsation.

In addition, there is a (relatively large) volume of the *non-pulsatile arterial blood* in the light path, which also contributes to the decrease in I. However, the extent to which the level of I is reduced by this absorber is *almost constant* over time; however, *respiratory activity* and changes in the *blood oxygenation* influence this absorber and thus contribute to relatively slow changes in I over time (sections "Respiratory Activity" in Sect. 5.1.2.3 and "Blood Oxygenation" in Sect. 5.1.2.3). Furthermore, the presence of both *capillary blood* and *venous blood* (Sect. 2.5.1) paired with their non-pulsatile nature contributes to the non-pulsatile decrease in the transmitted I. Capillary and venous blood induce an almost constant *offset in the absorption strength* (Fig. 5.14a). Lastly, *tissue* (almost bloodless, see section "Volume Effects" in Sect. 5.1.2.2) in the transilluminated region also contributes to the non-pulsatile absorption of light. All aforementioned non-pulsatile absorbers contribute to the *non-pulsatile direct component* I_{DC} of I (Fig. 5.14a). Thus the component I_{DC} is inversely related to the *direct part of the total absorbance.*

The *total transmitted intensity* I as a function of time t can be expressed as

$$I(t) = I_{AC}(t) + I_{DC}(t)\,. \tag{5.13}$$

The difference between the successive *peaks* and *troughs* of I represents the strength of the pulsatile absorber and thus corresponds to the *oscillation magnitude of* I_{AC} (Fig. 5.14a). Correspondingly, all non-pulsatile (and almost constant) absorbers account for the light absorption at the troughs of I and thus determine the size of I_{DC}.[38]

(Footnote 37 continued)
blood pressure yields a larger change in the venous volume than arterial volume; likewise, the volume pulse is amplified while the pressure pulse passes from stiff arterioles, through anastomoses, to compliant venules. The pulsatile venous volume can be expected only in tissue *regions close to arterio-venous-anastomoses*, i.e., veins located more proximally do not show venous pulses because the shunted arterial pulses become damped with increasing propagation distance (see pulse propagation in Sect. 2.5.2.3). To give an experimental example, the optical pulsatile deflection amplitude (comparable to $s_{S,D}$ in Fig. 5.15a) was shown to be proportional to the venous pulse pressure within the finger as the recording site (Kim 1986).

The arterio-venous-anastomoses are *sympathetically* controlled, i.e., sympathetic denervation opens these anastomoses. Consequently, the (reflex) vasoconstriction or stress-related reactions tend to reduce the optical pulsatile deflection amplitude because of closed anastomoses (closed shunts) and reduced pulsatile changes in the local *venous volume*. Finally, it should be noted that the arterio-venous-anastomoses are *abundant in the fingers*—to facilitate their *thermoregulatory functions* (Sect. 3.1.5)—in contrast to the *earlobe*; see Footnote 53 for more details.

[38] In fact, *alternating (pulsatile)* components in $I(t)$ can be *isolated* (relatively) easily from *direct (non-pulsatile)* components. For instance, the alternating intensity $I_{AC}(t)$ can be extracted out of

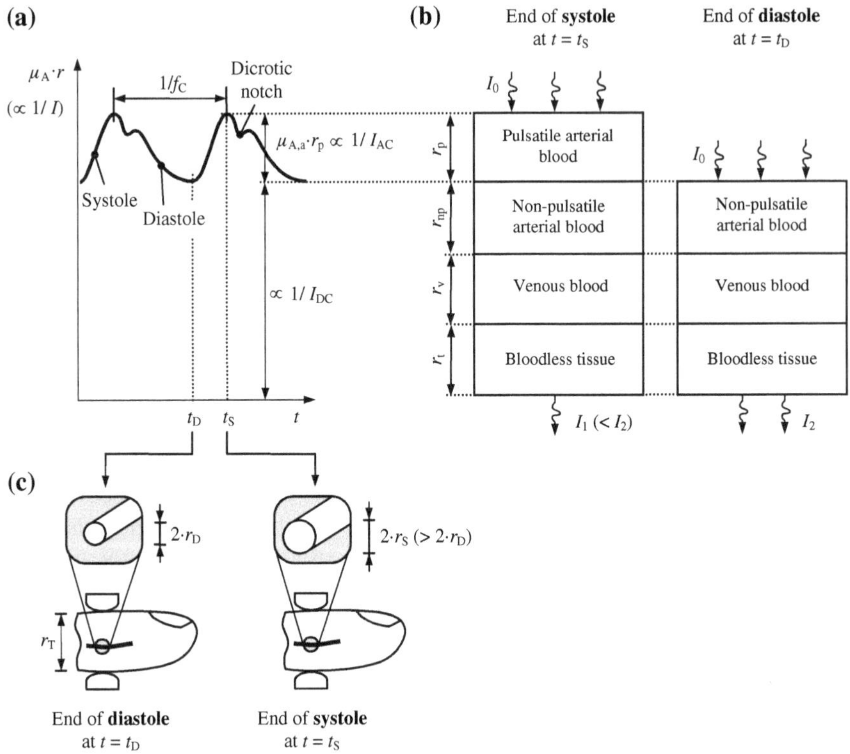

Fig. 5.14 The genesis of an induced optic biosignal. The incident light intensity I_0 mutates to the transmitted light intensity I (5.4) after passing living tissue. The intensity I ($I = I_{AC} + I_{DC}$) is composed out of the pulsatile component I_{AC} due to pulsatile arterial blood and, on the other hand, the non-pulsatile component I_{DC} due to non-pulsatile arterial blood, venous blood, and (bloodless) tissue. The respective path lengths of light are denoted by r with the total path length $r_T = r_t + r_v + r_{np} + r_p$; compare Fig. 5.6. (**a**) The light absorption $\mu_A \cdot r$ as a function of time with μ_A as the light absorption coefficient. (**b**) The corresponding compartmental model of living (perfused) tissue. (**c**) Blood perfusion in a finger for two time instances: end of diastole and end of systole (Fig. 3.17); compare Fig. 5.1 and Footnote 40. If only a single arterial vessel is assumed in the finger, the oscillation magnitude of r_p is given by $2 \cdot r_S - 2 \cdot r_D$ and $r_{np} = 2 \cdot r_D$, whereas r_S is the end-systolic artery radius and r_D the end-diastolic artery radius (Footnote 39)

The approximate *fractions* of the respective absorbing components are indicated in Fig. 5.15c. It is important to observe that the oscillation magnitude of I_{AC} *amounts to only about 5 % to the total I*, whereas *pulsatile and non-pulsatile blood* account for more than 50 % of the total light absorption.

From a *physiological point of view*, light (intensity) passing through biological tissue experiences a relatively *fast modulation* not only by

---

(Footnote 38 continued)
the total intensity $I(t)$ (5.13) with a *high-pass filter*, whereas the direct intensity $I_{DC}(t)$ can be extracted with a *low-pass filter*; compare Figs. 5.26 and 5.27.

- *cardiac activity* (Sect. 3.1.1) but also by other processes such as
- *respiratory activity* (Sect. 3.1.2) and *subject motions* (section "Motion Artefacts" in Sect. 5.1.2.3). A relatively *slow modulation* in the time domain occurs due to
- (slow) changes in *blood oxygenation* (Sect. 3.1.4).

Cardiac Activity

Pulsatile pressure waves—in response to *heart* activity—propagate along arterial vessels (arteries and arterioles but not veins, compare Footnote 37) yielding their local and temporal widening (Sect. 2.5.2.3). As illustrated in Fig. 5.14c, the local *arterial radius* (or local *arterial volume*) is at its minimum at the *end of diastole* and is at its maximum closely[39] before the *end of systole*[40]; compare Figs. 2.38 and 3.17. The incident light transversing an arterial vessel is temporarily attenuated in direct relation to the vessel's radius. Consequently, the transmitted intensity I, namely, the *pulsatile intensity component* I_{AC}, oscillates with f_C showing consecutive peaks and troughs (Fig. 5.14a). Here it should be noted that both *pulsatile changes of the arterial radius* in Fig. 5.14c and the corresponding deflection width of I_{AC} (related to the absolute level of I) in Fig. 5.14a are exaggerated for the sake of illustration. In fact, real changes of the radius from diastole to systole are relatively small and amount only up to 10 %; compare Fig. 3.17 and Sect. 2.5.1. Likewise, stronger *light absorption* and less received light intensity result during the (local) *blood surge* since each heart beat periodically increases the arterial blood volume in the transilluminated region (Fig. 5.1a, b).

In particular, the *light absorption* strength in the transilluminated region, namely, the product $\mu_A \cdot r$, *increases* with the influx of arterial blood *during systole*, i.e., local systole (Footnote 40); see also Figs. 5.14a and (5.4). The reverse is true during local diastole. For instance, Fig. 5.1a approximates the end of diastole and Fig. 5.1b

---

[39] It should be noted that the *arterial radius* reaches its *maximum value* in synchrony with the *maximum pressure* in the vessel in the course of its *pulsatile widening*. However, the maximum pressure can be expected closely before the end of the *local systole* (in peripheral arterial vessels, Footnote 40) because *systole ends* with the *dicrotic notch* and not with the maximum pressure; compare Fig. 2.38. The latter applies only for the *reflectionless propagation of pulses* (Sect. 2.5.2.3).

[40] The terms *systole* (ventricular contraction and ejection) and *diastole* (ventricular relaxation and filling), strictly speaking, are related to the activity of *heart ventricles* (Sect. 2.4.2). For instance, these terms are useful to interpret the pulse of the blood pressure leaving the heart along the aorta (Fig. 2.32b). In the case of *distal vessels* or even *peripheral arterial vessels* in a finger—as depicted in Fig. 5.14c—the terms *systole and diastole can be used only conditionally.*

In particular, the *pressure pulse* arrives at a *peripheral site* with a certain *time delay* (known as *pulse arrival time*, Sect. 3.1.3.1) which depends on both the *distance* from the heart to the peripheral site and the *velocity* of the propagating pulse (3.6). The local blood surge in the periphery—or the *local systole*—is delayed with respect to the *ventricular systole* in the heart; likewise, the decreasing blood volume in the periphery after the blood surge—or the *local diastole*—is also delayed with respect to the *ventricular diastole*.

Table 5.2 Qualitative and dominant modulation of the absorption coefficient μ_A of blood and the path length r of light through the volume of blood in accordance with Fig. 5.14b

Peripheral medium	Cardiac activity		Respiratory activity		Oxygenation changes	
	(local) systole	(local) diastole	Inspiration	Expiration	Deoxygenation	Oxygenation
Arterial blood	$r_p \uparrow$	$r_p \downarrow$	$\Delta r_p \downarrow r_{np} \downarrow$	$\Delta r_p \uparrow r_{np} \uparrow$	$\mu_A \uparrow$ (red light)	$\mu_A \downarrow$ (red light)
Venous blood	–	–	$r_v \downarrow$	$r_v \uparrow$		

These modulations determine the size of the product $\mu_A \cdot r$ and are considered for cardiac activity (Fig. 5.14a, c), respiratory activity (Fig. 5.15b, c), and changes in blood oxygenation (Fig. 5.8). Increasing $\mu_A \cdot r$ means increasing light absorption and decreasing transmitted light intensity (5.4)

approximates the end of systole, thus it becomes obvious that the effective (average) product $\mu_A \cdot r$ increases from Fig. 5.1a to Fig. 5.1b. This is because *during (local) systole there is an increasing amount of arterial blood* in the transilluminated region at the expense of the displaced surrounding tissue. In particular,

- μ_A of blood is significantly larger than μ_A of (almost bloodless) tissue (Table 5.1) and, on the other hand,
- the systolic radius r_S of a vessel is larger than its diastolic radius r_D (Fig. 5.14c).

Both circumstances facilitate an increase in the effective product $\mu_A \cdot r$ and in the bulk absorption of light by perfused tissue during systole. That is, the approximation $\mu_A \cdot r \approx \mu_{A,a} \cdot 2r_S$ applies for *systole* (at the time instance of the end of systole) and the approximation $\mu_A \cdot r \approx \mu_{A,a} \cdot 2r_D$ ($< \mu_{A,a} \cdot 2r_S$) applies for *diastole* (the end of diastole). Here $\mu_{A,a}$ denotes (almost *oxygenated*) *arterial blood*; see Fig. 5.8 with a typical *hemoglobin oxygen saturation S* in the range of 96–99 % and reflect on Fig. 3.19 and Footnote 20. That is, the light absorption by *oxyhemoglobin* determines $\mu_{A,a}$ because the amount of oxyhemoglobin in arterial blood is normally much higher than that of *deoxyhemoglobin*; the typical values of S in arterial blood are usually around 98 %.

Likewise, the product $\mu_{A,a} \cdot r_p$ *is inversely related to the instantaneous* $I_{AC}(t)$ with r_p *as the instantaneous path length* of light through the volume of the pulsatile arterial blood in tissue (Fig. 5.14b). As shown in Table 5.2, the path length r_p increases during systole and decreases during diastole. The *oscillation magnitude of* I_{AC} is determined by the term $\mu_{A,a} \cdot 2(r_S - r_D)$.

The *cardiac modulation* of the *light scattering* in the peripheral tissue seems to be of little importance. Table 5.1 reveals that μ_S' of blood and μ_S' of tissue do not differ, whereas g is slightly higher in blood than tissue. It means that during the blood surge, or the local increase in the blood volume (Fig. 5.1b), the effective (average) level of μ_S (5.7) increases a little. Thus, there is a tendency of light to be scatted a bit stronger in perfused tissue during *systole*.

The behaviour of the *induced optic biosignal*—as registered for diagnosis and therapy—is shown in Fig. 5.15a provided that *only cardiac activity* is present. Since this biosignal is (per definition) inversely related to the transmitted *light intensity* or, in other words, is proportional to the light *absorption strength* (Fig. 5.14a), the *biosignal exhibits*

- a relatively *steep systolic increase* and
- a *slow diastolic decrease* (Footnote 40).

Strictly speaking, this assumed inverse relationship between the transmitted light intensity and the light absorption strength applies only for small changes in the light absorption; see the differential equation in Footnote 19 and (5.4). In fact, the registration of *cardiac activity* by the optical sensor relies on the technology of the *optical plethysmography*; see Sect. 5.2.

The optic biosignal contains information about the *blood pressure* and *vascular compliance* in the transilluminated vascular beds. The *systolic-diastolic deflection* $s_{S,D}$ of the *optic biosignal optoplethysmogram* s_{OPG} (Fig. 5.15a) is proportional to the *pulsatile systolic-diastolic blood volume* $V_{S,D}$. In analogy with (2.23), this pulsatile volume is proportional to the product of the local systolic-diastolic deflection $p_{S,D}$ of *blood pressure* (section "Pulse Waveforms of Pressure and Flow" in Sect. 2.5.2.3) and the *arterial compliance* of the vascular wall (= V/κ with V as the total volume and κ the *module of volume elasticity* of the vessel, section "Pulse Propagation" in Sect. 2.5.2.3):

$$s_{S,D} \propto V_{S,D} \approx p_{S,D} \cdot \frac{V}{\kappa}. \tag{5.14}$$

In other words, the (local) deflection $s_{S,D}$ increases with increasing (local) $p_{S,D}$ and increasing (local) distensibility of the vessel. Unfortunately, the module κ (a measure for stiffness of the vessel, (2.23)) is not constant over the varying blood pressure (Sect. 2.5.2.1) so that $s_{S,D}$ is not a proportional measure to $p_{S,D}$ (5.14). In fact, the level of κ *rises with increasing blood pressure*; correspondingly, the arterial compliance decreases. As depicted in Fig. 5.16, the module κ is inversely proportional to the *local slope* of the *relationship between the vessel radius and blood pressure*; compare Fig. 2.42. This implies that $s_{S,D}$ progressively decreases with increasing *mean blood pressure*, given a constant $p_{S,D}$ within the vessel, as illustrated by the inequality $s^1_{S,D} > s^2_{S,D}$ from Fig. 5.16.

It should be noted that the *waveform of* s_{OPG} and the *waveform of blood pressure* are notably different due to this *non-linear relationship* from Fig. 5.16. Furthermore, the differences in the latter waveforms increase with increasing *mean blood pressure* because the *non-linear relationship* between the radius and pressure progressively saturates with increasing pressure. On the other hand, the latter waveforms become identical up to a scaling factor if the pressure deflection $p_{S,D}$ is relatively small. That is, a *linear relationship* between the radius and pressure can be assumed for a relatively *small deflection of blood pressure* (with κ assumed as constant).

An important feature of s_{OPG} is the presence of the *dicrotic notch* in its waveform, as illustrated in Fig. 5.14a. That is, the dicrotic notch in the waveform of *blood pressure* (Fig. 2.38) usually manifests in s_{OPG}, which allows for the determination of the transition from the local systole to the local diastole (Footnote 40). However,

- the *non-linear relationship* between the radius and pressure and, on the other hand,
- the *reflection phenomenon* of pulsatile waves in arterial vessels (Sect. 2.5.2.3)

may *shift, smooth, or highlight the dicrotic notch* in s_{OPG} (Figs. 3.36 and 5.32a) or even *introduce additional minima and maxima* in the *pulsatile component of* s_{OPG} (Fig. 2.53).

Besides the addressed *similarity* of the pulsatile waveform of s_{OPG} and that of *blood pressure*, these waveforms are *highly correlated* in different areas of the body such as finger and ear. In particular, these waveforms are mutually synchronized in

- the *0.1 Hz range* (10 s rhythm, see *Mayer waves* in Sect. 3.2.2) and in
- the *respiration range* at the respiratory rate f_R (usually $f_R > 0.1$ Hz); compare Sect. 3.2.3.

In addition, fluctuations of s_{OPG} and fluctuations of *interbeat intervals* (or of the instantaneous f_C) correlate with each other in both frequency ranges, i.e., in the 0.1 Hz range and f_R range (compare Figs. 5.33d and 5.31b); this is mainly because of the tight mutual interconnection between blood pressure and f_C (e.g., Fig. 3.40). It should be recalled that *low frequency oscillations* (around 0.1 Hz) are mainly related to *sympathetic activity* while *high frequency oscillations* (at the rate f_R) are mainly related to *parasympathetic activity*; for details see Sect. 3.1.1.

Interestingly, the waveform of s_{OPG} *is leading the waveform of blood pressure* in the *0.1 Hz region* (Bernardi 1996). The latter authors conclude that 0.1 Hz fluctuations in s_{OPG} were not passively transmitted to the skin (where the optical sensor is applied) from large arteries but instead 0.1 Hz fluctuations are *actively induced* by *sympathetic activity* of *smooth muscles* in the arterial wall. In contrast, s_{OPG} *was lagging behind blood pressure* in the *0.25 Hz region* at the fixed breathing rate $f_R = 0.25$ Hz (Bernardi 1996), showing that 0.25 Hz fluctuations in s_{OPG} were due to *passive mechanical transmission* of blood pressure fluctuations (caused by respiration itself, Sect. 3.2.2) to the transilluminated microvessels of the skin. Likewise, fluctuations of s_{OPG} synchronous with the respiration are mostly due to mechanical effects on the circulation and are not modified by changes in the sympathetic tone. The authors in Bernardi (1997) also report that these 0.1 Hz oscillations—unlike oscillations related to respiration—appear first in the *microcirculation* and later in the *(central) blood pressure*. It also indicates that these 0.1 Hz oscillations do not represent passive transmission from larger arteries but most likely originate in the microvasculature. In fact, both low frequency and high frequency oscillations decrease in their amplitude with *increasing age* because of progressive *loss of vascular distensibility* (Sect. 2.5.1).

Respiratory Activity

Besides *cardiac activity* (section "Cardiac Activity" in Sect. 5.1.2.3), respiratory activity modulates dynamically the (local) light absorption in multiple ways. In fact,

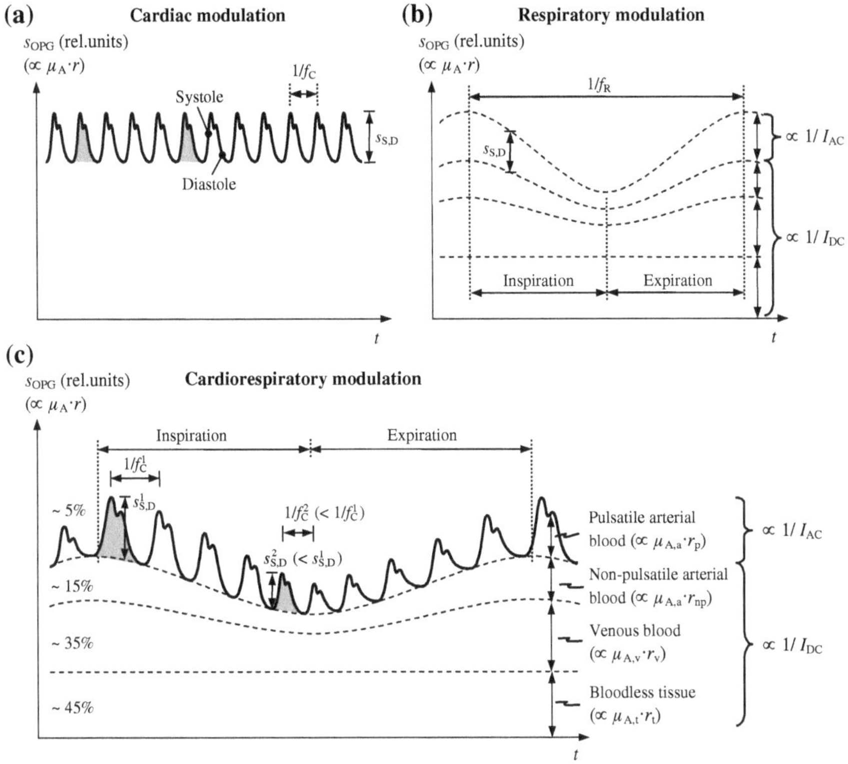

Fig. 5.15 Cardiorespiratory modulation of the light absorption in biological tissue as revealed by an induced optic biosignal optoplethysmogram s_{OPG}; compare Fig. 5.14. (**a**) Isolated cardiac modulation (pulsatile component) of s_{OPG} with indicated heart rate f_C, which shows increasing absorption during systole and decreasing absorption during diastole. (**b**) Isolated respiratory modulation of s_{OPG} with indicated respiratory rate f_R. (**c**) The resulting cardiorespiratory modulation with indicated approximate fractions of the light absorption due to arterial blood, venous blood, and tissue (Zourabian 2000). The systolic-diastolic deflection $s_{S,D}$ of s_{OPG} is indicated, which decreases during inspiration while f_C increases ($f_C^2 > f_C^1$)

respiration modulates all absorbing components of the *compartmental model of (peripheral) tissue* (Fig. 5.14b):

- pulsatile arterial blood,
- non-pulsatile arterial blood,
- venous blood, and
- bloodless tissue.

Consequently, both components the *pulsatile* I_{AC} *and non-pulsatile* I_{DC} *are affected by respiration*, as illustrated schematically in Fig. 5.15b and described below in detail. First, we consider the *physiological influence* of respiration on the respective absorbing components and then derive the corresponding change in I and the product $\mu_A \cdot r$ (5.4) as a measure of the *light absorbance*.

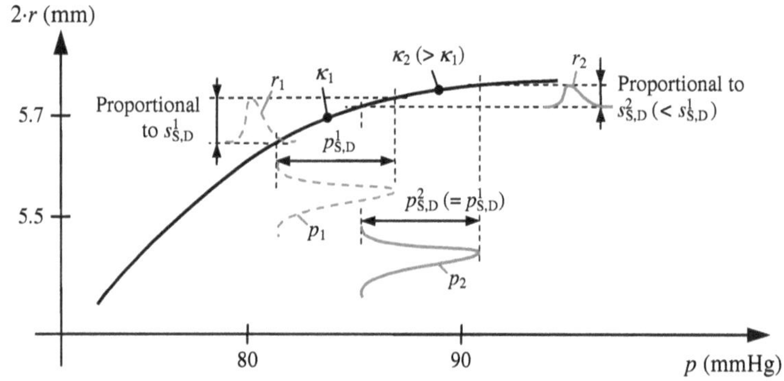

Fig. 5.16 Qualitative relationship of the vessel diameter $2 \cdot r$ and arterial blood pressure p within the vessel with the corresponding time courses. Approximate absolute values of r and p represent the human carotid artery in accordance with Fig. 2.42. The waveform of p_2 exhibits a higher mean pressure than that of p_1, whereas the corresponding systolic-diastolic deflections $p_{S,D}$ are equal

The amount of the *pulsatile arterial blood* decreases during *inspiration* (Fig. 5.15b) due to slightly *decreased left ventricular stroke volume*, i.e., the volume of blood ejected with each heart beat. Numerous physiologic phenomena contribute to the inspiratory decrease of this stroke volume; namely, mechanisms of *respiratory pump*, *ventricular interdependence*, *respiratory sinus arrhythmia*, and *reverse thoracic pump*. The latter mechanisms are described in section "Normal Respiration" in Sect. 3.2.1.1 (and Footnote 227 in Sect. 3) in detail. Consider also mechanisms counteracting a temporal unbalance of the left ventricular stroke volume (Sect. 3.2.2.1) in order to attain a complete picture of the body behaviour over the respiration cycle.

In consequence of the decreased left ventricular stroke volume during inspiration, the *pulsatile change of the arterial circumference* decreases. The systolic-diastolic deflection of the *arterial radius* (or arterial volume) also decreases and thus the *pulsatile attenuation* of light in tissue. That is, the oscillation magnitudes of both the *pulsatile* I_{AC} *and* $\mu_{A,a} \cdot r_p$ *decrease during inspiration*, as shown in Fig. 5.15b, c. Obviously the reverse is true during *expiration*, as summarized in Table 5.2.

The amount of the *non-pulsatile arterial blood* decreases slightly during *inspiration* (Fig. 5.15b) due to temporarily *reduced left ventricular output*, i.e., the total volume of blood pumped by the heart over a particular period of time. In fact, the left ventricular stroke volume is tightly interrelated with the ventricular output (via 2.30). A momentarily decreased stroke volume necessarily means a momentarily decreased output provided that f_C is constant. However, as *time passes*, regulating mechanisms begin to (partly) compensate for the reduced left ventricular output while *increasing* f_C; for details see Sect. 3.2.1.1. It should be noted that the respiration-induced *oscillation of* f_C *always follows* (or lags behind) that of *blood pressure* which is tightly interrelated with the *left ventricular stroke volume*; for details see Sect. 3.2.2.1. Figure 5.15c illustrates (delayed) increasing f_C in the course of inspiration.

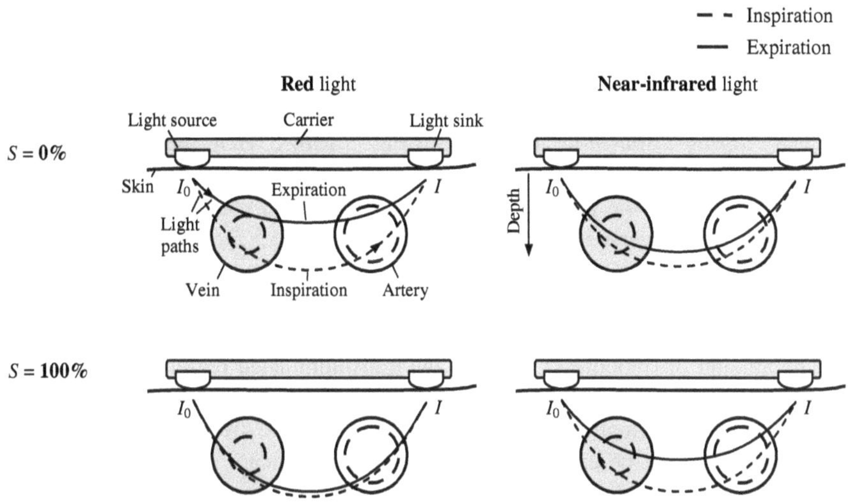

Fig. 5.17 Respiratory modulation of the light absorption in biological tissue using the reflectance mode (Fig. 5.22). The average (non-pulsatile) cross-sections of peripheral veins and arteries decrease during inspiration (Fig. 3.28). This yields a corresponding increase in the transmitted light intensity I given a constant incident intensity I_0 (Fig. 5.15c). In addition, decreased blood volume during inspiration tends to increase the propagation distance of light in tissue and thus the probing depth of light (Fig. 5.20). The absolute depth and its relative increase during inspiration are inversely related to each other and are mostly determined by the light absorption in blood. The absorption strength in blood is a strong function of the hemoglobin oxygen saturation S and the light colour (Fig. 5.8)

In consequence of the temporarily *reduced left ventricular output* during *inspiration*, the blood filling of (peripheral) arteries diminishes, thereby reducing the *non-pulsatile arterial circumference*, as illustrated in Fig. 5.17. Likewise, the mean peripheral perfusion or the non-pulsatile (peripheral) arterial volume decreases, which reduces the attenuation of the incident light and contributes to *increase of the non-pulsatile* I_{DC}. Therefore the term $\mu_{A,a} \cdot r_{np}$ *decreases during inspiration*, as shown in Fig. 5.15b, c, with r_{np} as the path length of light through the volume of the non-pulsatile arterial blood (Fig. 5.14b). The reverse is true during expiration. Table 5.2 signifies the decrease of r_{np} during inspiration and its increase during expiration.

The amount of the *peripheral venous blood* also decreases during *inspiration* due to increased pressure gradient between the peripheral veins (exposed to atmospheric pressure) and intrathoracic veins (subatmospheric pressure, compare Footnote 117 in Sect. 2). This causes venous blood to be drawn from the peripheral veins into (large) intrathoracic venous vessels, according to the mechanism of *respiratory pump* (Sect. 3.2.1.1). Consequently, the *vein circumference* in the periphery decreases strongly during inspiration. As depicted in Fig. 5.17, this decrease of the vein circumference is much more pronounced than the aforementioned decrease of the non-pulsatile *arterial circumference* during inspiration.

This is because the *venous compliance* is much larger than the *arterial compliance* (Sect. 2.5.1). Thus, the volume of the peripheral venous blood decreases during inspiration, in the course of which light in tissue is attenuated less by venous blood. This contributes to *increase of the non-pulsatile* I_{DC} and to *decrease of the product* $\mu_{A,v} \cdot r_v$ *during inspiration*, as shown in Fig. 5.15b, c. Here the coefficient $\mu_{A,v}$ reflects (partly *deoxygenated*) *venous blood* (see Fig. 5.8 with 40 % < S < 75 % and Fig. 3.19) and r_v is the path length of light through the volume of venous blood (Fig. 5.14b). The reverse is true during expiration (Table 5.2). Figure 5.17 indicates also diverging light paths for varying oxygenation of blood and varying colour of the incident light; the corresponding phenomena are discussed in section "Specific Issues" in Sect. 5.1.2.3 (and Footnote 44).

The amount of (bloodless) *tissue* in the transilluminated region remains almost constant over the respiration cycle. In fact, the optical sensor in the *transmittance mode* (Fig. 5.22)—usually applied in *distal regions* (on extremities) such as finger or earlobe—can not be expected to register any significant changes in the *tissue fraction of the non-pulsatile* I_{DC} in synchrony with *breathing* (Fig. 5.15b, c). That is, *tissue deformations* determining this tissue fraction and occurring with f_R are not pronounced in distal regions of the body.

In contrast to distal regions, *proximal regions* such as the chest region (especially, regions close to the lungs) are subjected to noticeable (superficial) tissue deformations during *respiration*. Here the optical sensor on the (chest) *skin* is usually operated in the *reflectance mode*, as illustrated in Figs. 5.17 and 5.22. Tissue deformations yield versatile effects which affect *directly* the tissue fraction of I_{DC} and *indirectly* all other arterial and venous fractions (from the compartmental model of tissue, Fig. 5.14b):

- During *inspiration*, the *chest skin and its adipose layer* are *stretched* to a certain extent so that imbedded vessels (especially, compliant venous vessels, compare section "Motion Artefacts" in Sect. 5.1.2.3) are slightly squeezed and their (mean) circumference tends to decrease. Consequently, this would contribute to
  - a slight decrease of the *arterial fraction* of the *pulsatile* I_{AC}, i.e., decrease of the oscillation magnitude of I_{AC};
  - a slight increase of the *arterial fraction* of the *non-pulsatile* I_{DC}; and
  - increase of the *venous fraction* of the *non-pulsatile* I_{DC}; compare Fig. 5.15b, c. In addition,
- the *elastic dermis* of the *skin* and the adjacent *adipose layer* become thinner in the course of their stretching *during inspiration*. The *reduced thickness* of these layers tends to reduce the light attenuation by the skin and thus may allow a deeper *probing of light* below the skin than during expiration, given the *reflectance mode*. Underlying tissues such as skeletal *muscles* may predominantly be probed during inspiration. Consequently, this would contribute to a slight increase in the total light absorption because the muscle absorbs more strongly the incident light than the *skin* or the *adipose layer* (below the skin); compare the corresponding size of μ_A from Table 5.1 and Fig. 5.7. Therefore,

  - a slight increase in the *pulsatile* I_{AC} can be expected with the *muscle* being probed because here the fraction of the probed pulsatile *arterial blood* (modulating I_{AC} by definition) is relatively high. On the other hand,
  - a slight decrease of the *non-pulsatile* I_{DC} can also be expected because the total probed *non-pulsatile blood* volume in the *muscle* (modulating I_{DC} with blood as the principle absorber) is higher than in the *skin* or *adipose tissue*. In particular, it can be expected that the above mechanism related to the muscle probing is relevant only if a relatively thick adipose layer is present under the skin, as being applicable to the chest skin of *obese patients*.

- Provided the *reflectance mode* is applied on a proximal site such as the *chest skin*, the sensor carrier (with the attached light source and sink, Fig. 5.17) is rather *flexible* and *prebent* when applied on the convex skin with an adhesive band, then the *alignment of the light source and sink* relative to one another can vary over the *respiration cycle*. In particular, *during inspiration* the source and sink would slightly move away from each other because the chest circumference increases; likewise, the skin is locally flattened. The *effective path length of light* in tissue increases, which yields a corresponding decrease of the *tissue fraction of* I_{DC}. However, since the *source-sink distance* (of about 1 cm) is very small in relation to the *chest circumference* (of about 80 cm) and, on the other hand, the relative respiration-induced change in the circumference is usually less than 10 %, the corresponding alignment of the source and sink is very small and thus the corresponding change in I_{DC} can usually be neglected.

As discussed above, the *tissue fraction of the non-pulsatile* I_{DC} can not be expected to show respiration-induced changes in *distal skin regions*. The light absorption in *proximal skin regions*, in contrast, is sensitive to respiration-induced *tissue deformations* which affect not only the tissue fraction of I_{DC} but also, indirectly, arterial and venous fractions of I_{AC} and I_{DC} (Fig. 5.14b). However, the associated changes of I_{AC} and I_{DC} with the respiration cycle mostly counteract each other so that unidirectional tendencies in these changes are missing. Correspondingly, Fig. 5.15b, c assumes absent changes in the *tissue fraction of* I_{DC} over the respiration cycle. Likewise, the (absorption) product $\mu_{A,t} \cdot r_t$ can be assumed to be constant in a first approximation. Here the coefficient $\mu_{A,t}$ reflects the light *absorption by (bloodless) tissue* (Table 5.1) and r_t the path length of light through the volume of tissue (Fig. 5.14b).

In summary, the *light absorption* is modulated by respiration in *multiple ways*, as demonstrated in Fig. 5.15c. All components of the *compartmental model of tissue* (Fig. 5.14b) experience periodic changes with f_R, as already discussed in detail. It should be noted that the depicted sinusoidal behaviour of the light absorption in Fig. 5.15c is only an approximation. In short, the oscillation magnitude of the *pulsatile component* I_{AC} of I temporarily decreases during *inspiration* while the *non-pulsatile component* I_{DC} of I temporarily increases due to the discussed circumference changes in arterial and venous vessels. It seems that the *respiratory component* is mostly *representative within the venous fraction of* I_{DC} or within the appropriate product $\mu_{A,v} \cdot r_v$ (Franceschini 2002). This owes to the large *venous compliance* related to the limited *arterial compliance*. In addition, the rate of cardiac pulses in I_{AC}, i.e., the instantaneous frequency f_C, increases with inspiration, following the phenomenon of *respiratory sinus arrhythmia* (see section "Normal Respiration" in Sect. 3.2.1.1). Lastly, the absolute *width of the pulse wave* of the optic biosignal varies over the respiratory cycle. This variation is related to both changes in f_C (due to respiratory sinus arrhythmia) and morphological changes in the *pulse waveform* (due to respiration-induced changes in blood pressure and *reflected waves*, Fig. 3.27); compare Figs. 5.16 and 3.36a.

Finally, it should be noted that the discussed *respiration-induced tissue deformation* is *similar* to the *motion-induced tissue deformation* in terms of their impact on the light absorption changes. For instance, the inspiratory stretch of a *proximal* skin region predominantly squeezes venous vessels (less pressurized, highly compliant vessels) embedded within the skin, which displaces local *venous blood* and thus induces a corresponding change in the *non-pulsatile* I_{DC} (Fig. 5.15b, c). In analogy, a movement of a *distal* extremity such as finger flexion yields local deformation of tissues (in the finger) in which *venous blood* is also easily displaced. According to section "Motion Artefacts" in Sect. 5.1.2.3, the light absorption by venous blood experiences motion-induced changes, contributing to the temporal variability of the *non-pulsatile* I_{DC}. Paradoxically, (local) absorption changes due to (local) tissue deformations—induced by either the respiration or patient motion—are usually considered as *artefacts* in the registered optic biosignals (section "Motion Artefacts" in Sect. 5.1.2.3). This is because these artefacts—which in fact provide useful physiological data—usually impair the *accuracy of the monitoring of* S (section "Specific Issues" in Sect. 5.1.2.3).

Blood Oxygenation

After the optic behaviour of tissue has been discussed within the scope of *cardiac* and *respiratory activity* (sections "Cardiac Activity" in Sect. 5.1.2.3 and "Respiratory Activity" in Sect. 5.1.2.3), the effects of *blood oxygenation* on the light absorption in living tissue should be discussed. *Oxygenation changes* are typically *slow* in the time domain (Sect. 3.1.4), which provides a rather *slow modulation* of the light absorption (Fig. 5.1a, c) in comparison with its relatively *fast modulation* by cardiac and respiratory activity (Fig. 5.1a, b).

General Issues

Usually the level of the *hemoglobin oxygen saturation* S in arterial blood is of *diagnostic interest*. That is, the absorbance of *arterial blood* should be separated from that of *venous blood* and (bloodless) *tissue*. This is where the *pulsation of arterial blood*—unlike venous blood and other tissues (compare Footnote 37)—can be used as a distinguishing feature of arterial blood.

The *compartmental model of tissue*—as shown in Fig. 5.14b—offers a solid and conclusive basis to establish a *model for the experimental estimation of* S. As already discussed, the compartmental model describes the physiological origin of the pulsatile (alternating) component I_{AC} and non-pulsatile (direct) component I_{DC} in the transmitted light intensity I (Sect. 5.1.2.3). As a special feature of this experimental estimation of S, the *pulsation of arterial blood* will explicitly be used. In fact, this technology using pulsations is coined as *pulse oximetry*,[41] i.e., the registration of S is based on two technologies *spectrometry* and *optical plethysmography*; for details see Sect. 5.2.

Following the compartmental model of tissue (Fig. 5.14b), the *absorption law* from (5.4) can be applied to each absorbing component and rewritten as

$$\begin{aligned} &\ln\left(\frac{I}{I_0}\right) = -\mu_A \cdot r = -\left(\mu_{A,a} \cdot r_p + \mu_{A,a} \cdot r_{np} + \mu_{A,v} \cdot r_v + \mu_{A,t} \cdot r_t\right) \quad \text{or} \\ &I = I_0 \cdot \left(e^{-\mu_{A,a} \cdot r_p}\right) \cdot \left(e^{-\mu_{A,a} \cdot r_{np}}\right) \cdot \left(e^{-\mu_{A,v} \cdot r_v}\right) \cdot \left(e^{-\mu_{A,t} \cdot r_t}\right). \end{aligned} \tag{5.15}$$

---

[41] *First oximeters* to assess S *of arterial blood* had only *one wavelength* in the *red region* of spectrum (Zourabian 2000). As illustrated in Fig. 5.8, μ_A of oxyhemoglobin in this spectral region is markedly different from μ_A of deoxyhemoglobin. However, the single wavelength application prevented this oximeter to assess the total hemoglobin concentration and thus to compensate for possible changes in the total hemoglobin concentration (Sect. 3.1.4). A reference measurement (or a *calibration* procedure) was necessary on *bloodless tissue* without the *primary chromophore* (section "Volume Effects" in Sect. 5.1.2.2), e.g., on tissue made temporarily bloodless by its squeezing (see below).

Later oximeters with two wavelengths were introduced. They used one *red wavelength* usually around 660 nm as a *measure for oxyhemoglobin* and a second *near-infrared wavelength* at around 800 nm as a *reference* (at the *isosbestic point*, Fig. 5.8).

Today's oximeters use the second wavelength at higher near-infrared wavelengths at around 890 nm (section "Light Wavelength" in Sect. 5.2.1.2). Here μ_A of oxyhemoglobin and μ_A of deoxyhemoglobin show *reverse courses* over the wavelength with respect to the *isosbestic point.* That is, μ_A at red light decreases with increasing *oxygenation* of hemoglobin while μ_A at near-infrared light correspondingly increases (Fig. 5.19). This reverse behaviour is explicitly used in the assessment of S (section "Blood Oxygenation" in Sect. 5.1.2.3). In addition, the *pulsatile nature of the transmitted light intensity* is also used today in the estimation of S in order to focus on oxygenation of *pulsatile arterial blood* only (Footnote 49). The latter technology is known as *pulse oximetry.*

As already noted, *early oximeters* used a *calibration procedure*, in which

- tissue was *compressed* to eliminate blood (Kamat 2002). The light absorbance by *bloodless tissue* ($\propto I_{DC}$, compare with the compartmental model of tissue, Fig. 5.14b) was used as a baseline to *isolate arterial blood* and to estimate S of this arterial blood (especially from $I - I_{DC}$, see 5.13). In addition,
- *pneumatic cuffs* were introduced to measure increase in the light intensity when, for instance, the transilluminated ear was squeezed (compare sections "Motion Artefacts" in Sect. 5.1.2.3 and "Contacting Force and Skin Temperature" in Sect. 5.2.1.2). It was also customary to
- *heat tissues* (e.g., earlobe) in order to filter out absorption effects due to venous and capillary blood. The (local) skin heating is known to produce (local) *vasodilation* of vascular beds under the skin surface, in the course of which the *pulsatile component* of the transmitted light intensity increases (Mendelson 1988); compare Fig. 3.21. For instance, the latter authors observed a five-fold increase in the *pulsatile ratio* $\mathcal{R}$ (5.17) by increasing the local skin temperature from 34 to 45 °C, whereas the optical sensor was applied on the forearm and operated in the *reflectance mode* (Fig. 5.22b).

Here $\mu_{A,a}$ denotes the absorption coefficient of oxygenated arterial blood, $\mu_{A,v}$ the absorption coefficient of deoxygenated venous blood (Fig. 5.8), and $\mu_{A,t}$ the absorption coefficient of bloodless tissue (Table 5.1); compare Fig. 5.15c.

Provided that (small) pulsations of blood alter only the size of r_p (Fig. 5.14b), the *oscillation magnitude of the pulsatile component* I_{AC} can be obtained from (5.15) by the derivative of the total I with respect to r_p. After the derivatives dI and dr_p are approximated by differences ΔI and Δr_p, respectively, the latter derivative of I with respect to r_p yields

$$\Delta I = -I_0 \cdot \mu_{A,a} \cdot e^{-(\mu_{A,a} \cdot r_p + \mu_{A,a} \cdot r_{np} + \mu_{A,v} \cdot r_v + \mu_{A,t} \cdot r_t)} \cdot \Delta r_p \approx I_{AC} \,. \tag{5.16}$$

Then the *pulsatile ratio* $\mathcal{R}$ of I_{AC} to I_{DC} (i.e., the pulsatile intensity *normalized* to non-pulsatile intensity) can be established on the assumption that I_{DC} is almost equal to I (compare (5.13)). This assumption is justified by the experimental observation that I_{AC} amounts to only 1–5 % of the total I (Fig. 5.15c). The ratio $\mathcal{R}$ can be expressed as

$$\mathcal{R} = \frac{I_{AC}}{I_{DC}} \approx \frac{I_{AC}}{I} = -\mu_{A,a} \cdot \Delta r_p \,. \tag{5.17}$$

It should be noted that the above ratio $\mathcal{R}$—in contrast to I_{AC} and I_{DC} ((5.15) and (5.16))—is no longer a function of the incident I_0, the amount of the non-pulsatile absorbers in tissue, and even the sensitivity of the light sink.

Since we are interested in the oxygenation of arterial blood characterized by its *coefficient* $\mu_{A,a}$ and the size of the ratio $\mathcal{R}$ can be experimentally assessed, the unknown parameter Δr_p in (5.17) is to be eliminated in the following procedure. Please note that the *minus sign* in (5.17) denotes decreasing I ($\Delta I < 0$, see (5.16)) with increasing path length r_p ($\Delta r_p > 0$); the true consequence of the light absorption in tissue.

This can be done by using *two wavelengths of light*: *red light* and *near-infrared light* (Fig. 5.8); compare Footnote 41. It should be noted that *red and near-infrared light* reside within the *optical window* into tissue (Fig. 5.7) in order to penetrate tissue to (relatively) large depths and even transverse distal extremities such as the finger. In other words, red and near-infrared light readily penetrate biological tissues while green and blue light (with relatively short wavelengths) as well as infrared light (with relatively long wavelengths) are strongly absorbed by these tissues; see section "Volume Effects" in Sect. 5.1.2.2.

Therefore, another *ratio* R of $\mathcal{R}_R$ at red light to $\mathcal{R}_{IR}$ at near-infrared light can be calculated, to give

$$R = \frac{\mathcal{R}_R}{\mathcal{R}_{IR}} = \frac{I_{AC}^{R} / I_{DC}^{R}}{I_{AC}^{IR} / I_{DC}^{IR}} \approx \frac{\mu_{A,a}^{R}}{\mu_{A,a}^{IR}} \,. \tag{5.18}$$

Here the *superscript* R *and* IR denote red and near-infrared light, respectively. The above approximation—yielding the ratio of $\mu_{A,a}$ at red light to $\mu_{A,a}$ at near-infrared light—*assumes identical* Δr_p for both light colours (compare Figs. 5.17 and 5.20).

It is important to observe that the ratio R can be experimentally measured and R is a function of $\mu_{A,a}$ only. That is, the ratio R does not depend on hardly accessible values of I_0, r_p, and other parameters of the non-pulsatile absorbers.

The relationship between the saturation S and the absorption coefficient $\mu_{A,a}$ should be now established in order to *estimate* S while *measuring* R (5.18). In fact, the coefficient $\mu_{A,a}$ (compare (5.4)) can be expressed as

$$\mu_{A,a} = \rho \cdot \sigma = \frac{H}{V} \cdot (S \cdot \sigma_{HbO} + (1 - S) \cdot \sigma_{Hb}). \tag{5.19}$$

Here H is the hematocrit[42] (about 0.45), V the volume of a single red blood cell (about 90 μm^3), σ_{HbO} and σ_{Hb} the respective absorption *cross section of the red blood cell* containing totally *oxygenated hemoglobin* (S = 100 %) and totally (reduced) *deoxygenated hemoglobin* (S = 0 %). That is, the *number density* ρ in (5.19) is approximated as the ratio H/V, whereas σ_{HbO} and σ_{Hb} are weighted by S and $(1 - S)$, respectively (Schmitt 1991), according to the definition of S in (3.12).

Combining (5.18) and (5.19), the *relationship between S and R*—a *model for the experimental estimation of S*—can be easily derived:

$$S = \frac{R \cdot \sigma_{Hb}^{IR} - \sigma_{Hb}^{R}}{(\sigma_{HbO}^{R} - \sigma_{Hb}^{R}) + R \cdot (\sigma_{Hb}^{IR} - \sigma_{HbO}^{IR})}. \tag{5.20}$$

Even though all *cross sections* (of the red blood cell) in the above equation are *constants*, the above *relationship* between S and R is highly *non-linear*. In particular, the linearity of this relationship increases with decreasing difference between σ_{Hb} and σ_{HbO} for near-infrared light.

Equation (5.20) transforms to an (almost[43]) *linear relationship* if the *isosbestic point* (in Fig. 5.8) is chosen for near-infrared light at about 800 nm. At this point, the cross sections σ_{Hb} and σ_{HbO} are equal at near-infrared light and (5.20) simplifies to

$$S = \frac{R \cdot \sigma_{Hb}^{IR} - \sigma_{Hb}^{R}}{\sigma_{HbO}^{R} - \sigma_{Hb}^{R}} = c_1 \cdot R + c_2 \approx -0.28 \cdot R + 1.12\,. \tag{5.21}$$

Here c_1 and c_2 are constants determined only by cross sections for the different oxygenation states (i.e., S = 0 and 100 %) and different light colours (i.e., 660 nm for red light and 800 nm for near-infrared light). The constant c_1 represents the *slope* of this relationship (the curve of S over R) while c_2 represents the *offset*. In particular, the given numerical values of c_1 and c_2 were estimated from $\sigma = \mu_A/\rho$ with μ_A extracted from Fig. 5.8 and ρ estimated from the ratio H/V (5.19);

---

42 The *hematocrit* is the volume percentage of blood occupied by the (packed) red blood cells.

43 Strictly speaking, the *relationship between S and R*—as given in (5.21)—would become *linear* only if $c_2 = 0$. This is because the linearity requires that two mathematical properties, namely, *homogeneity and additivity*, are satisfied. In fact, they are satisfied only if $c_2 = 0$.

i.e., $\mu_A = 0.771$ and $0.0846\ \text{mm}^{-1}$ at 660 nm for $S = 0$ and 100 %, respectively, whereas $\mu_A = 0.196\ \text{mm}^{-1}$ at 800 nm in the *isosbestic point*.

It can be observed that the derived model for the *estimation of S* from the *experimental ratio R* shows an *inverse relationship between S and R* because the slope $c_1 < 0$ in (5.21). Figure 5.18 demonstrates this inverse relationship. The estimation of S is possible only *once per heart beat* (in a first approximation (Wukitsch 1988)) because the amplitude of the *pulsatile deflection is needed* as input parameter (5.18).

Figure 5.19 illustrates the *pulsatile behaviour of the optic biosignal* for the different oxygenation states of blood and different colours of the incident light. For *low values of S* and *red light*, the *pulsatile deflection* is relatively large because the relevant μ_A *of blood is relatively high* (Fig. 5.8). In fact, the higher is μ_A of blood, the stronger is the pulsatile change in the (local) *light absorbance* from systole to diastole (section "Cardiac Activity" in Sect. 5.1.2.3). In analogy, the higher is μ_A of blood, the larger is the (total) decrease in the component $I_{AC}(t)$ during (local) *systole* and the subsequent (total) increase during (local) *diastole* (compare Fig. 5.15a). A similar behaviour is demonstrated in Fig. 5.1a, c. The red light is absorbed less by oxygenated blood in the shown vessel (Fig. 5.1a) than deoxygenated blood (Fig. 5.1c).

For *low values of S* and *near-infrared light* (Fig. 5.19), the *pulsatile deflection* of the optic biosignal is relatively small because the relevant μ_A *of blood is relatively low* (Fig. 5.8). Provided the discussed pulsatile deflections, it follows from (5.18) that the *ratio R is relatively high for low S*, which actually confirms the derived inverse relationship from (5.21) (Fig. 5.18).

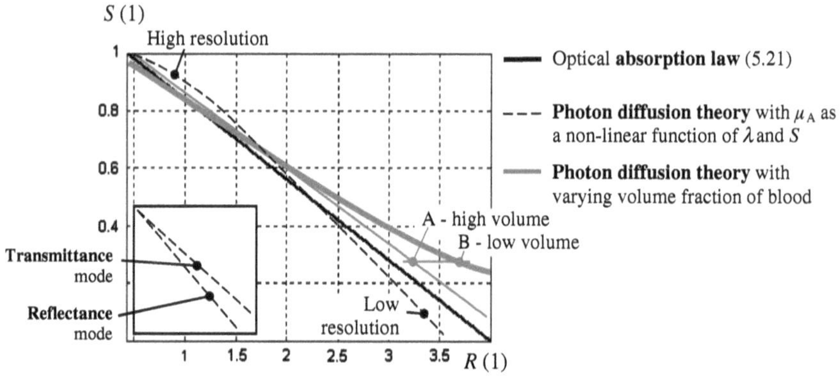

Fig. 5.18 The hemoglobin oxygen saturation S as a function of the experimental ratio R (following the absorption law, 5.21) with near-infrared light located at the isosbestic point (Fig. 5.8). Effects of varying blood volume (e.g., 1 and 5 %) in tissue and the non-linear wavelength λ dependence of the absorption coefficient μ_A are indicated, as predicted by the photon diffusion theory (5.22–5.25). The *embedded subfigure* demonstrates the calibration curves of the transmittance and reflectance modes for comparison aims (at $r_T = r_R$, see Fig. 5.22)

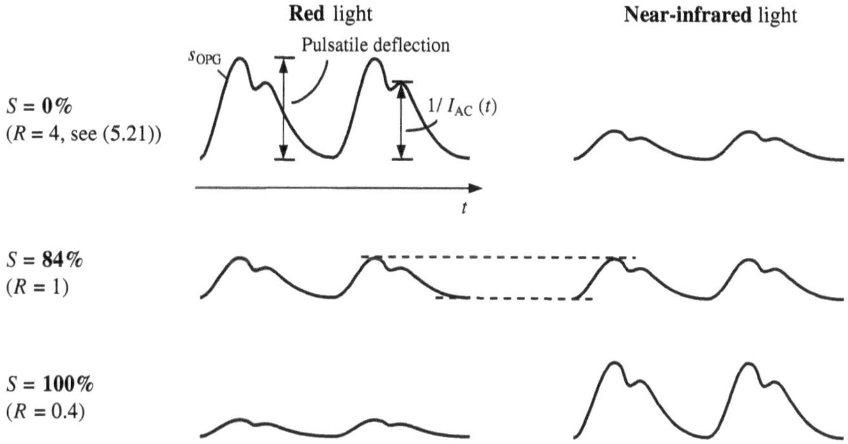

Fig. 5.19 Oscillatory changes of the pulsatile component $I_{AC}(t)$ of the transmitted light intensity as revealed by an induced optic biosignal optoplethysmogram s_{OPG}; compare Fig. 5.14a. These changes are shown in relation to different levels of the hemoglobin oxygen saturation S for red and near-infrared light; compare Fig. 5.8

With *increasing S*—as shown in Fig. 5.19—the magnitudes of the pulsatile deflections for red and near-infrared light become increasingly similar. For instance, the saturation *S = 84 % yields the ratio R = 1*, which results in (almost) equal amounts of the absorbed red and near-infrared light. For *high values of S* close to 100 %, in contrast to low values of S, the pulsatile deflection is relatively small for red light and large for near-infrared light because of the corresponding behaviour of μ_A (from Fig. 5.8). Since the deflection width depends on the size of μ_A, it is obvious that the highest deflection is given for low S and red light, the next lower deflection for high S and near-infrared light, and the smallest deflection for high S and red light.

Specific Issues

The derived *model for the experimental estimation of S*—as summarized by (5.21)—is subject of numerous *limitations*. In fact, only the *absorption law* was considered (5.15) and applied on a simple *compartmental model of tissue* (Fig. 5.14b). Important effects have been neglected such as

- relatively strong *scattering* of light in biological tissue (section "Inhomogeneity Effects" in Sect. 5.1.2.2);
- *different pathways* of red and near-infrared light in tissue with the optical sensor operated in the *reflectance mode* (Fig. 5.22b), which yields different (average) lengths of the paths travelled by photons of both wavelengths (Fig. 5.20);
- *diffuse reflections* of light in the heterogeneous tissue (Fig. 5.13);

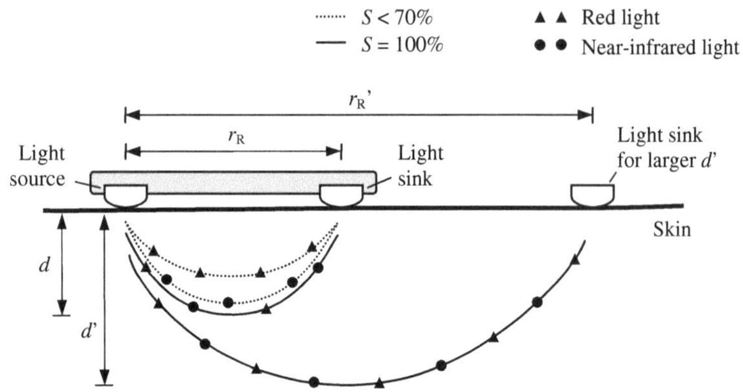

Fig. 5.20 Mean probing depth d as a function of the varying source-sink distance r_R in the reflectance mode (Fig. 5.22). Light penetration paths are indicated in tissue below the skin for red and near-infrared light and varying hemoglobin oxygen saturation S; compare Fig. 5.17

- *geometrical dimensions* of the optical sensor in the transmittance and reflectance modes, especially the distance between the light source and sink (Fig. 5.22);
- presence of the *direct light* in the *reflectance mode*, i.e., the light from the light source to the sink without passing through tissue (section "General Issues" in Sect. 5.2.1.2); and
- *pulsatile signal components* within I due to
  - other hemoglobin derivatives in arterial blood such as methemoglobin and carboxyhemoglobin, in addition to reduced and oxygenated hemoglobin (Footnote 220 in Sect. 3),
  - pulsatile changes in the bulk scattering of transilluminated tissue,
  - pulsatile changes in the local thickness of bloodless tissue around vessels, and
  - (minor) venous pulsations (compare Footnote 37).

The *photon diffusion theory*—as introduced in section "Inhomogeneity Effects" in Sect. 5.1.2.2—can be applied to account for pulsatile changes in the *bulk absorption coefficient* of perfused tissue, *multiple scattering* effects (pulsatile changes in the bulk scattering are neglected), *dissimilar optical path lengths* of red and near-infrared photons in tissue, and the actual *distance between the light source and sink* (Schmitt 1991). In short, this theory is applicable if scattering greatly exceeds absorption and the light sink is many *diffusion lengths* away from the light source (i.e., $D \ll r_T$, r_R, compare (5.9) and Fig. 5.22).

The *photon diffusion theory* yields the *relationship between S and R*, i.e., another *model for the experimental estimation of S*, a model which is similar to that derived from the absorption law (5.20):

$$S = \frac{R \cdot \sigma_{\mathrm{Hb}}^{\mathrm{IR}} - c \cdot \sigma_{\mathrm{Hb}}^{\mathrm{R}}}{c \cdot \left(\sigma_{\mathrm{HbO}}^{\mathrm{R}} - \sigma_{\mathrm{Hb}}^{\mathrm{R}}\right) + R \cdot \left(\sigma_{\mathrm{Hb}}^{\mathrm{IR}} - \sigma_{\mathrm{HbO}}^{\mathrm{IR}}\right)} \tag{5.22}$$

with

$$c = \frac{\left(\mu_{\mathrm{S}}'\right)^{\mathrm{R}} \cdot \left(c'\right)^{\mathrm{R}}}{\left(\mu_{\mathrm{S}}'\right)^{\mathrm{IR}} \cdot \left(c'\right)^{\mathrm{IR}}}. \tag{5.23}$$

Here c' is a function of the *total attenuation coefficient* α of perfused tissue and of the *geometrical dimensions* of the light propagation path in tissue (Fig. 5.22). Likewise, the function c' is not independent on S and the incident light *wavelength* because μ_{A} and, consequently, α (5.10) are affected by changes in S and changes in the wavelength (Fig. 5.8). Tissue is treated as a *semi-infinite homogenous medium* with uniform absorption and scattering throughout tissue.

In the *transmittance mode*, the function c' results to

$$c' = \frac{\alpha \cdot r_{\mathrm{T}} - 1}{\alpha^2 \cdot r_{\mathrm{T}}}, \tag{5.24}$$

where r_{T} *is the distance between the light source and sink* (*along the light beam*, Fig. 5.22a). The above equation applies only if $\alpha \cdot r_{\mathrm{T}} \gg 1$, i.e., the transilluminated tissue is sufficiently thick in comparison with the light *penetration depth* $1/\alpha$ (5.10).

In the *reflectance mode*, the function c' results to

$$c' = \frac{-r_{\mathrm{R}}^2}{1 + \alpha \cdot r_{\mathrm{R}}}. \tag{5.25}$$

The above equation applies only provided that the *distance* r_{R} *between the light source and sink* (*perpendicular to the incident light beam*, Fig. 5.22b) is sufficiently large and the tissue thickness r_{T}' below the skin is semi-infinite in extent, i.e., $\alpha \cdot r_{\mathrm{T}}' > 2\alpha \cdot r_{\mathrm{R}} \gg 1$.

In comparison with the simple *relationship between S and R* (5.20)—which is derived from Beer-Lambert *absorption law* and includes only σ_{Hb} and σ_{HbO} as parameters—the *photon diffusion theory* yields a similar but more sophisticated relationship (5.22). In addition to the *absorption* characteristics σ_{Hb} and σ_{HbO} of tissue, (5.22) includes *scattering characteristics* of tissue and *geometrical characteristics* of the optical sensor. In particular, the function c describes effects of *dissimilar optical path lengths*[44] at the two wavelengths and thus effects of *dissimilar total attenuation* of light in perfused tissue.

Today's *pulse oximetry*—as used for *diagnostic purposes*—is mainly based on the simplified relationship from (5.21). The *experimental ratio* R (5.18) is calculated out of the transmitted light intensities I_{AC} and I_{DC} (see Footnote 38), which then is used as the only input parameter of (5.21) in order to *estimate the saturation* S. Usually, the values of c_1 *and* c_2 in (5.21) are not assumed as constants but instead are *experimentally determined* for pre-defined levels of S (in the clinically relevant range of usually between 70 and 100 %) and the corresponding measured levels of R. Moreover, different optical sensors require different c_1 and c_2 (as functions of S) because the distance r_T (or r_R) may vary from one sensor to another (Fig. 5.22); compare (5.24) and (5.25). This experimental evaluation of c_1 and c_2 eliminates some shortcomings of (5.21), especially those related to *multiple scattering* effects and *dissimilar optical path lengths* at the two wavelengths (Schmitt 1991). The resulting *experimental relationship between* S *and* R, namely, the values of c_1 and c_2, is stored as the *calibration curve* within the diagnostic device, namely, the *pulse oximeter*.

In practice, the *accuracy and resolution* of the assessment of S strongly depends on the effectiveness of the *calibration curve*[45] which, at best, specifies the most

---

[44] The *dissimilar propagation distances* of *red* and *near-infrared photons* (before they enter the light sink) occur due to the wavelength dependence of μ_A and μ_S'. For instance, for *low values of* S and *red light*, the relevant μ_A *of blood is relatively high* (Fig. 5.8) which determines strong absorption of red light. In consequence, the average *propagation distance of red light* (photons) in tissue is decreased for low S. It reduces the *sensitivity* of the corresponding I to (local) absorbance changes in the *deeper layers of tissue* relative to the sensitivity measured with *near-infrared light* (Schmitt 1991). For this reason, the (local) *slope of* S *versus* R curve—as shown in Fig. 5.18—increases for *low* S (< 80 %), thereby decreasing the effective *resolution of* ΔS for a given ΔR. For high S (> 80 %), the reverse is true yielding relatively high resolution of S. Figure 5.20 illustrates decreased propagation distance and decreased *probing depth* of *red light* for low S in comparison with *near-infrared light*. Similar observations can be derived from Fig. 5.17 while comparing light paths during *expiration*; e.g., red light for low S is confined to shallow depths.

In contrast to the *photon diffusion theory*, both *propagation distances* at the two wavelengths are assumed to be equal and do not depend on optical properties of tissue in the model of *Beer-Lambert absorption law* (5.4).

[45] A reliable *testing and calibration of oximeters* is an important issue. Usually, *in-vivo* empirical calibrations on *volunteers* are performed, having the volunteers to inspire hypoxic gas and analysing their arterial blood gas samples as reference. The reduced partial pressure of oxygen in the inspired air (Footnote 219 in Sect. 3) decreases the resulting blood oxygenation level; such calibration is usually performed in a stepwise manner from $S = 100$ % down to only 70 % (Venema 2012). Also animal *tissues with active oxygenation* can be used for calibration purposes (Zonios 2004). *In-vitro* calibrations can be carried out by the use of heparinised *blood samples* (Reichelt 2008).

Besides blood as calibrating media, *phantom media* (artificial media) can also be used for testing and calibration of oximeters. For instance, *scattering and absorption properties* of tissue can be experimentally approximated by using suspended polystyrene microspheres, e.g., with 1 μm diameter, the approximate size of biological cells (Mourant 1998; Kumar 1997). Calibrated intralipid solutions, special rubber and resin can also be used to mimic *scattering tissues*, whereas an ink is added to simulate *chromophores* in absorbing tissue (Benaron 2005). In addition, patented solutions exist to imitate *oxygenation changes* of blood using electrically-controlled light absorption based on artificial media such as liquid crystals (Winter 2002).

probable S value associated with the measured R value (provided a *typical set of physiological conditions*, see below). In particular,

- the *accuracy of the calibration curve* is proportional to the *linearity* of this curve over the relevant range of S, whereas
- the *resolution of the calibration curve* is inversely proportional to the *slope of S versus R* curve (i.e., the slope c_1 from (5.21), compare Footnote 44). That is, the smaller the change ΔS for a given ΔR, the larger is the resolution.

The following discussion is devoted to the *accuracy and resolution* aspects of the *experimental assessment of S* using *transmittance* and *reflectance modes* in the sensor application (Fig. 5.22). Here numerous *physiological circumstances* and physical effects are involved:

Variations in Physiological Parameters

There is a composite effect of *normal variations* in *physiological parameters* such as *tissue blood volume, hematocrit level, finger thickness, hemoglobin concentration,* which all affect the relationship between S and R and can vary among subjects and even in the same subject over recording time. Obviously, the necessary calibration procedure to attain the *calibration curve* is influenced by the *population of subjects* participating in it, considering inevitable physiological *variations among subjects*. In fact, the *accuracy* of the calibration curve can be assessed in an analytic way using the *photon diffusion theory* (Schmitt 1991); compare (5.22–5.25). That is, this theory shows that the accuracy error is less than a few percent for *high* $S > 70$ % on the assumption that the aforementioned physiological parameters are subjected to *normal fluctuations* (among subjects). The *accuracy* decreases significantly for *low* $S < 70$ %, i.e., the variance of the calibration curve increases. For instance, the maximum error in S shows a disproportional increase from 3 % for $S = 70$ % up to 20 % for $S = 20$ %. This error mainly increases due to prevailing effects of *multiple scattering* and, on the other hand, the wavelength dependence of μ_A and μ_S' in tissue.

For instance, the sensitivity of the calibration curve to a variation in *finger thickness*, i.e., a variation in the distance r_T (or r_R, Fig. 5.22), increases significantly for *low S* and *low blood volume* in tissue (Schmitt 1991). Namely, rising r_T (or r_R) increases the *slope of S over R* curve, thereby decreasing the *resolution* of the calibration curve.

Effects of μ_A

Non-linear dependence of μ_A on the light wavelength λ and S is an important issue; compare Fig. 5.8. That is, the *accuracy and resolution*, i.e., the linearity and slope of the calibration curve, respectively, are strongly influenced by this non-linear dependence (Schmitt 1991). In general, it is clear from (5.21) that the larger is the *difference between* σ_{Hb} *and* σ_{HbO} for red light, the smaller is the slope c_1—or the *slope of S over R* curve—and thus the better is the *resolution* of the calibration curve.

According to Footnote 44, the sensitivity of *red light* to μ_A changes in tissue is decreased for low S (< 80 %) relative to the sensitivity of *near-infrared light*. This is because the corresponding *propagation distance* of red light (average *path length* of red photons) in tissue is reduced (Fig. 5.20). In consequence, the (local) *slope of S versus R* curve increases for *low S*, thereby decreasing the (local) resolution, as illustrated in Fig. 5.18. The reverse is true for high S (> 80 %) where the calibration curve yields relatively high resolution.

Effects of μ_S'

Non-linear dependence of μ_S' (and μ_S) on λ and S is another important issue. As described in section "Inhomogeneity Effects" in Sect. 5.1.2.2, the scattering intensity, and so the coefficient μ_S', increase with decreasing λ. As a result, *red light* is scattered stronger than *near-infrared light*. The intense scattering reduces the effective *propagation distance* (or the *probing depth*, compare (5.26)) of red light in tissue. In analogy to Footnote 44, the reduced light *propagation distance* decreases the sensitivity of red light to μ_A changes in tissue. Likewise, the intense scattering tends to increase the (local) *slope of S versus R* curve, thereby decreasing the (local) resolution. This *limited resolution* is especially prominent for *low S*; see the effect of the non-linear dependence of μ_A on λ and S from above.

Varying Blood Volume

Varying blood volume in tissue, i.e., the particular *volume fraction of blood* in tissue (of about 5 %, section "Volume Effects" in Sect. 5.1.2.2), affects the *slope and offset of the calibration curve*; this being especially relevant for patients with anemia or ischemia (Schmitt 1991; Mannheimer 1997). An increase in the blood volume at low $S < 70$ % causes increased *optical absorption of red light relative to that of near-infrared light*. Again by analogy with Footnote 44, the *propagation distance* of red light decreases, which yields an increase in the (local) *slope of S versus R curve*, as illustrated in Fig. 5.18. The *linearity* of the calibration curve seems to advantageously increase with rising blood volume (Schmitt 1991); compare Fig. 5.18. That is, *increased blood volume decreases the resolution* and *increases the accuracy* of the calibration curve. In other words, the ratio R (5.18) decreases for a given value of S provided an increased blood volume and *low S* in tissue; compare points A (high blood volume, e.g., 5 %) and B (low blood volume, e.g., 1 %) in Fig. 5.18. On the other hand, increased blood volume at *high S* has the opposite effect on the change in R for a given value of S, albeit to a lesser extent (Fig. 5.18).

Figure 5.17 illustrates *shortening of light paths* in tissue with *increasing local blood volume*; i.e., from inspiration to expiration (section "Respiratory Activity" in Sect. 5.1.2.3). The decreasing *propagation distance* reduces also the *probing depth* of light. It should be noted that the absolute depth at expiration and its relative increase during inspiration are inversely related to each other. This is because the higher is μ_A of blood (for given S and λ, Fig. 5.8), the stronger is the light absorption and thus the shallower is the light probing in tissue. For instance, at expiration the minimum depth occurs for *low S and red light*, whereas the maximum depth occurs for *high S and red light*. Consequently, the higher is the applicable μ_A of blood, the stronger is the *respiration-induced change* in the *light absorption* and thus the *respiration-induced change* in the *probing depth*; e.g., when the transilluminated *blood volume* becomes smaller in the course of inspiration. Therefore, the maximum increase in the probing depth with *inspiration* occurs again for *low S and red light*, whereas its minimum increase occurs for *high S and red light*.

For instance, the reduction of *R* at low *S* due to increased blood volume can be minimized by moving the *red wavelength* to the far red, e.g., 660 to 735 nm (Mannheimer 1997). Then the overall *slope of S versus R* curve becomes (disadvantageously) larger due to smaller difference between σ_{Hb} and σ_{HbO} for red light (i.e., the slope c_1 from (5.21) increases; compare Fig. 5.8), whereas the slope sensitivity to varying blood volume in tissue (advantageously) diminishes. In summary, *hypoperfusion, hypotension, and vasoconstriction*[46] strongly impede the accuracy of oximeter (Hill 2000).

Obviously *weak pulsations* of the blood volume (i.e., *low pulsatile fractions*, see section "General Issues" in Sect. 5.2.1.2) due to *poor perfusion* corrupt the correct estimation of the pulsatile deflection amplitude (Fig. 5.19) and thus the calculation of $\mathcal{R}$, *R* (see (5.17) and (5.18)) and the corresponding estimation of *S*. In particular, poor perfusion interrelated with low blood flow may result in *large motion artefacts* (see section "Motion Artefacts"in Sect. 5.1.2.3). In addition, *venous pulsations* (Footnote 37) may also impair the accuracy of the assessment of *S*.

An advantageous *combination of the transmittance mode with the reflectance mode* (Fig. 5.22) within a single optical sensor applied on a finger was reported to improve the estimation of *S* in states of *vasoconstriction* and *poorly perfused tissues* (Shafique 2012); compare Footnote 46. The authors built a (weighted) sum of the transmitted light intensities resulting from the transmittance and reflectance modes; then this sum was used to estimate *R* (5.18). In addition, the *reflectance mode* seems to be most easily compromised by poor perfusion. In this sense, the *transmittance mode* is inferior to the aforementioned combination of both modes (Shafique 2012). The low robustness of the reflectance mode could be justified by a superficial light probing combined with (particularly) strong decrease of the local peripheral perfusion in the superficial tissue layers in the case of *vasoconstriction*. A relatively *small*

---

[46] The *accuracy of oximeter* is low in patients with *decreased blood volume* (see above) and *diminished peripheral pulsation* (essential to calculate *R*, (5.18)). In fact, the accuracy of the *optical sensor applied on a peripheral site* is low in patients who need them most, such as

- patients with *hypotension*,
- cold extremities (*hypothermia*, compare Fig. 5.28), or
- *peripheral vasoconstriction*.

High *sympathetic tone* of such patients reduces significantly vascular pulsations in the periphery; compare Footnote 37. For instance, continuous intraoperative assessment of blood oxygenation during *anaesthesia* shows a relatively *low accuracy* (or requires special measures to reach an acceptable accuracy) because of—in part—*poor tissue perfusion* due to (intended) hypothermia and/or hypotension.

pulsatile fraction $\mathcal{R}$ pertaining to the *reflectance mode* (see section "General Issues" in Sect. 5.2.1.2), i.e., a relatively weak arterial pulsation in the superficial tissue, already indicates the least robustness of this mode to decreased blood perfusion.

Dissimilar Optical Path Lengths

Dissimilar optical path lengths of *red* and *near-infrared* photons in tissue play an important role. As discussed above, the (average) photon path lengths diverge at the applied wavelengths (Footnote 44, Figs. 5.17 and 5.20). An *optimal selection of* λ with respect to the *accuracy and resolution* of oximeter is an important issue because of the wavelength dependence of μ_A and μ_S' of perfused tissue (Mannheimer 1997; Schmitt 1991).

In fact, the wavelength dependence of μ_A and μ_S' in *blood determines these dissimilar path lengths* and the wavelength dependence of the *slope* of S versus R curve. The *bloodless tissue*, in contrast, contributes little to the wavelength dependence of the (average) path lengths because its properties μ_A and μ_S' show only weak dependence on λ in the relevant region of the spectrum (i.e., within the *optical window* into tissue) (Schmitt 1991).

It is interesting to note that the *calibration curve* is most stable and repeatable if the *fractional change in photon path length* in response to any *perturbation* in perfused tissue, e.g., due to *varying blood volume*, is equivalent at both λ (Mannheimer 1997). The effective optical *path lengths* (or the probing depth, Sect. 5.2.1.1) need to be well matched at the two λ, especially in the *reflectance mode* (Mannheimer 1997; Schmitt 1991).

This match of the *probing depths* was investigated by authors in Mannheimer (1997) using numerical Monte Carlo simulation[47] of the light propagation within a semi-infinite homogenous tissue with the isotropic scattering only. The authors suggest conventional wavelengths 660 and 900 nm *at high S* and optimised wavelengths 735 and 890 nm *at low S* to insure identical probing depths in the whole range of S. However, this improvement at low S—compare with the effects of the varying blood volume from above—does not come without a cost, the *slope* of S versus R curve becomes larger. In addition, the estimation of the *probing depth* from (5.26) indicates a match of the probing depths at the two λ provided that the corresponding products $\mu_A \cdot \mu_S'$ are equal.

Comparing the *transmittance and reflectance mode* (Fig. 5.22), it is obvious that the corresponding *paths of photons* of a given λ travelling from the light source to sink are different even if the distances r_T and r_R are equal. On average, the path in the reflectance mode is longer because of the loss of shallow penetrating photons at

---

[47] *Monte Carlo* simulation is based on modelling of possible photon trajectories, i.e., random walk of photons, from the light source through biological tissue to the light sink (compare Fig. 5.20). The histories of individual photons are simulated as they undergo *multiple absorption and scattering events*. Each photon is followed until it disappears in tissue or reaches the light sink. *Probability distributions* are used to describe propagation of photons. Around 1 million photons are usually used in such simulations (Meglinski 2002).

the (skin) surface. Consequently, the *calibration curves* of both modes differ substantially even for $r_T = r_R$ (e.g., 7 mm each), i.e., the transmittance mode shows a smaller *slope of S versus R* curve than the reflectance mode (Schmitt 1991), indicating a better resolution of the transmittance mode. Figure 5.18 demonstrates schematically these calibration curves for $r_T = r_R$. Interestingly, these curves tend to overlap for $r_T > r_R$ (e.g., 11 and 7 mm) because an increased r_T compensates for the dissimilar path length (Sect. 5.2.1.2). It is important to note that this simple relationship between the transmittance and reflectance mode (Fig. 5.18) is valid only in a *homogenous medium* and may not be valid for real *skin* with (highly) unevenly distributed blood within and below the skin.

Stability of Light Emission

Stability of light emission in the *light source*, i.e., a small change in the emitted λ or in the spectral width of the incident light (Fig. 5.8), affects substantially the *slope and offset of the calibration curve*. In particular, the stability of *red light* is more important than that of *near-infrared light*. This can be explained by strong changes of μ_A during blood (de)oxygenation given red light (e.g., at 660 nm almost tenfold change of μ_A from $S = 0$ to 100 %, see Fig. 5.8) in comparison to near-infrared light (at 890 nm only one and a half times change of μ_A from $S = 0$ to 100 %, see Fig. 5.8). A small change in λ of red light produces a substantial *offset* of the *calibration curve*, whereas this curve is much less sensitive to a change in λ of near-infrared light (Schmitt 1991). As a result, the stability of the emission spectrum of red light is an important factor influencing the *long-term stability of oximeter*.

Hemoglobin Derivatives

Other hemoglobin derivatives in blood can also impact the assessment of *S*. Since *methemoglobin* (Footnote 220 in Sect. 3) absorbs equal amounts of *red* and *near-infrared light* (at about 660 and 890 nm, respectively (Wukitsch 1988)), a high concentration of methemoglobin causes *R* to converge to 1 and thus *S* to approach 84 % given a typical calibration curve (5.21); compare Fig. 5.19. That is, the oximeter's reading stays at 84 %, irrespective of the true blood oxygenation (Kamat 2002). Another hemoglobin derivative, *carboxyhemoglobin* (Footnote 220 in Sect. 3), absorbs red light as does oxyhemoglobin (in totally oxygenated blood) and absorbs near-infrared light to a very little extent. Thus a significant concentration of carboxyhemoglobin will cause oximeter to overestimate the actual amount of oxyhemoglobin in blood and thus the effective level of *S* (Kamat 2002).

Accuracy and Resolution

Summing up the accuracy and resolution aspects, high *accuracy and resolution* in the assessment of *S* are given only for *high* $S > 70$ *%* with the accuracy being then in the range of a *few percent*. For *low S*, the wavelength dependence of μ_A and μ_S' and, on the other hand, varying blood volume at the measurement site strongly influence the *calibration curve*. An *optimization of the incident* λ yields an improvement in the *accuracy* for low S but does not come without cost; the *resolution* of the calibration curve decreases. There is a *tradeoff between the accuracy and resolution* of the calibration curve, i.e., a small slope of S versus R curve inevitably reduces the curve's linearity. From an engineering point of view, low resolution in combination with high accuracy introduces problems in the hardware design since accurate measurements of the experimental ratio R become more critical.

Time Delay

It should be noted that there may be a significant time delay (*response time*) between a change in the *blood oxygenation in the lungs* (as induced by e.g., holding breath or sudden decrease in inspired fractional oxygen concentration) and a corresponding change in the oximeter's reading S with the *optical sensor* applied *on the skin*. The following physiological phenomena and the applied methodology of signal processing contribute to this time delay:

- The response time is strongly related to the *application region* of the optical sensor (Kaniusas 2006). The more *proximally* (centrally) the sensor is applied, the smaller is the *time delay* in the registered S (Bebout 2001). Authors in (Jubran 1999) review a much faster response time to a sudden decrease in inspired oxygen concentration for the *ear probe* than the *finger probe* of about 10–20 s and 24–35 s, respectively. In addition, the time delay increases strongly in *distal* application regions with *poor perfusion* (decreased blood flow), *peripheral vasoconstriction*, or *hypothermia*. To give an example, hypoxemia manifests approximately 90 s later for the finger versus forehead in the case of peripheral vasoconstriction (Bebout 2001). Likewise, arterial oxygen in tissue may drop to dangerous levels before an alarm can be raised by *oximeter*.
- Standard *oximeters usually average* the registered values of S over the time interval of 5–20 s in order to reduce (dynamic motion) *artefacts* (Hill 2000); compare section "Motion Artefacts" in Sect. 5.1.2.3. In particular, dynamic motion artefacts in the (raw) optic biosignal are usually indistinguishable from normal *heart-induced pulsations* in this biosignal (unless e.g., electrocardiogram with its R wave is used for identification of cardiac pulses in the optical biosignal, compare Fig. 5.32a, b). Lengthening the averaging time increases the likelihood that there are more true cardiac pulses than motion artefacts in the averaging interval. However, increased averaging time delays the detection of (potential) hypoxemia and omits short transient desaturations. In general, such *averaging procedures* contribute to the aforementioned *time delay*.

Motion Artefacts

Lastly, *motion artefacts* may significantly impair the accuracy of oximeter or even lead to its *failure in the assessment of S*. As discussed in section "Motion Artefacts" in Sect. 5.1.2.3, *erroneous desaturations* result from motion artefacts, whereas *low blood perfusion* at the site of measurement reinforces even stronger *motion-induced desaturations* and thus increases failure rates even more.

Motion Artefacts

Motion of extremities or body parts close to or including the application region of the optical sensor induce *motion artefacts*; namely, artefacts in the effective *light absorption* and thus artefacts in the *optic biosignal* as registered by the sensor (Fig. 5.1). For instance, finger, wrist, or elbow movements are typical motions. In fact, different motion artefacts can be identified, namely,

- (periodic) *motion of the optical sensor* along the skin displacing the transilluminated region under the skin and thus inducing changes in the *local light absorption*, as registered by the sensor;
- (periodic) *motions of the optical sensor* perpendicular to the skin surface, thereby yielding changes in the *contacting force* (between the sensor and its application site on the skin) and thus changes in the local light absorption; and
- (periodic) *motion of body parts displacing venous blood* under the skin and thus inducing changes in the local light absorption (the optical sensor is immovable with respect to the skin).

A potential *motion of the optical sensor along the skin* dislocates the transilluminated region. Provided that tissue under the skin is highly *heterogeneous* in its structure and optical properties (Fig. 5.9a), the dislocation of light paths in tissue yields different absorption and scattering characteristics applicable to light. Consequently, the registered optic biosignal undergoes motion-induced changes, i.e., exhibits *motion artefacts*. In addition, the motion of the optical sensor may severely compromise the *coupling of light* from the light source into tissue as well as the coupling of light out of tissue into the light sink. Furthermore, *directional properties* of the light source and light sink amplify these motion-induced changes in the optic biosignal, especially when the optical sensor is operated in the *reflectance mode*; see section "General Issues" in Sect. 5.2.1.2.

Motions of the optical sensor perpendicular to the skin surface yield changes in the contacting force between the sensor and its application site on the skin. As discussed in section "Contacting Force and Skin Temperature" in Sect. 5.2.1.2 in detail, the varying *contacting force* impacts mechanically *arterial and venous vessels* below the skin surface and thus the light absorption by transilluminated arterial and venous blood. Figure 5.27 illustrates these motion-induced effects of the contacting force on the pulsatile and non-pulsatile components of the optic biosignal.

Local movements of body parts (e.g., during *exercise testing*) close to or including the application region of the optical sensor (stationary and immovable with respect to the skin) induce temporal fluctuations in the light absorption. Such fluctuations in the absorption are attributed to the *non-pulsatile absorbers* in tissue (Fig. 5.14b) and are mainly due to local displacements of *venous blood*. It should be stressed that venous blood amounts to about 80 % of the total blood volume (Sect. 2.5.1), whereas *blood* is the *principle absorber* in tissue (section "Volume Effects" in Sect. 5.1.2.2). The displacement of venous blood dominates because the intraluminal pressure within venous vessels is much lower than that within arterial vessels (Fig. 2.39). Consequently, a movement of an extremity such as a *finger flexion* yields local tissue deformation which, in turn, can easily displace *less pressurized venous blood* in this tissue but not arterial blood. The resulting *motion-induced changes* in the light absorption (by displaced *venous blood*) are usually indistinguishable from *heart-induced changes* in the light absorption (by *pulsatile arterial blood*).

Likewise, the arterial pulsation is less susceptive to external body motions than the volume of venous blood and cutaneous pulsations of capillaries. With the data in Sect. 2.5.1, Fig. 2.41b, and Asada (2003), it can be estimated that

- the *local volume of venous vessels* sustain *external pressures* up to about *15 mmHg*,
- *cutaneous pulsations* of *capillaries* sustain up to about *25 mmHg*, and
- *pulsations of arterial vessels* sustain up to about *80 mmHg* (i.e., approximately up to the level of diastolic blood pressure).

Consequently, the *reflectance mode* can be expected to be more sensitive to the external pressure than the *transmittance mode* (Table 5.3). This is because the light in the reflectance mode probes predominantly peripheral *superficial capillary beds* (section "General Issues" in Sect. 5.2.1.2) which are highly *susceptible to this external pressure.*

As illustrated in Fig. 5.21, *motion artefacts* obscure *optic biosignals.* Here the index finger on the right hand was temporarily subjected to dynamic motion, namely, consecutive extension and flexion at the rate of about 2–3 Hz. An optical sensor operated in the transmittance mode was applied on this finger in a way that the sensor could hardly be displaced by finger movements. The corresponding optoplethysmogram is shown in Fig. 5.21b. Before and after the *movement's period,* the pulsatile waves can be observed which oscillate with the *heart rate* f_C. In contrast, during the movements the waveform of the *optoplethysmogram becomes obscured* while the resulting *oscillation rate* of this waveform corresponds rather to the movement's rate but not to f_C. It can be expected that the observed motion artefacts are due to *displaced venous blood,* as discussed above.

Another optoplethysmogram from a *stationary index finger* on the left hand was recorded for *reference purposes.* As depicted in Fig. 5.21c, pulsatile waves can only be observed here, which occur with the rate f_C even during the movements period. Likewise, motion artefacts are missing within this reference optoplethysmogram.

More importantly, motion artefacts compromise not only optic biosignals but also the *accuracy of oximeter,* and usually increase its *response time* (section "Specific Issues" in Sect. 5.1.2.3). The effective *blood oxygenation* is usually underestimated in the course of motion artefacts, i.e., oximeter displays *erroneous desaturations during motions.* The *interference of motions* with the estimation of blood oxygenation is notably strong when the motion rate and the pulsatile rate f_C are similar. In this case, *pulsatile fractions* $\mathcal{R}$ of the transmitted light at the red and

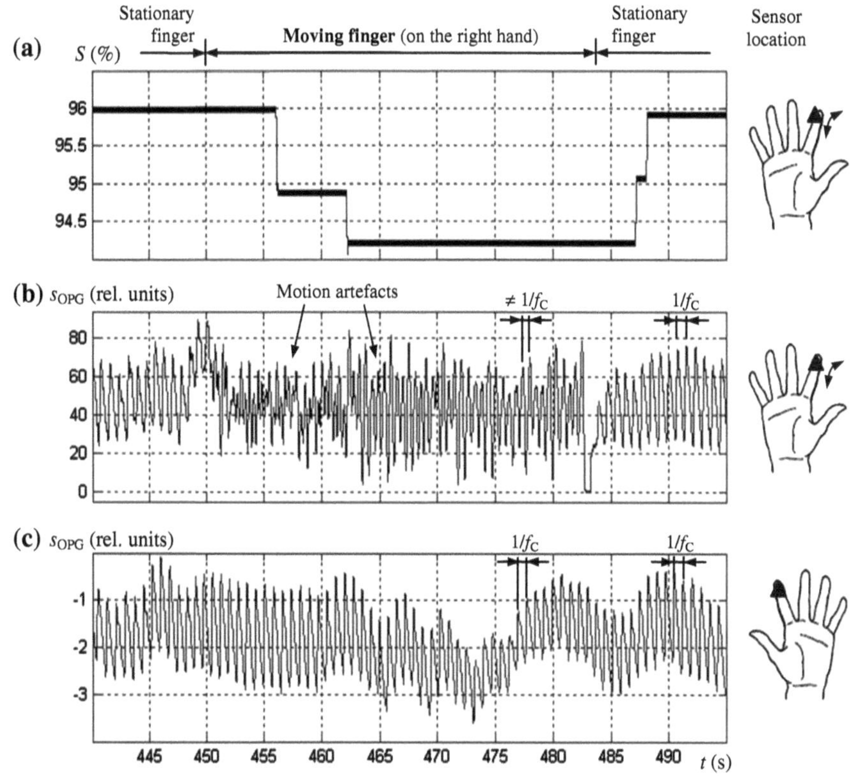

Fig. 5.21 Experimental effects of finger movements on the optic biosignal. (**a**) Hemoglobin oxygen saturation S derived from optic biosignals (from the moving index finger on the right hand) with the optical sensor operated in the transmittance mode (Fig. 5.22a). (**b**) Optic biosignal optoplethysmogram s_{OPG} (from the moving index finger on the right hand) with the same optical sensor as used in (**a**). The instantaneous heart rate f_C is indicated. (**c**) Another optoplethysmogram s_{OPG} (from a stationary fingertip on the left hand) with the optical sensor operated in the reflectance mode (Fig. 5.22b). Only the index finger on the right hand was temporarily subjected to a dynamic motion, i.e., consecutive extension and flexion at the rate of about 2–3 Hz

near-infrared light—as the only input parameters for the estimation of blood oxygenation (5.17, 5.18, 5.21)—become corrupted.

In particular, motion artefacts induce excessive *motion-induced deflections* (or noisy "pulsations") in the affected optoplethysmogram; compare Fig. 5.21b, c. These artificial deflections may even *overwhelm cardiac pulsations* with f_C (Goldman 2000), as shown in Fig. 5.21b. Consequently, the fractions $\mathcal{R}$ (5.17) for both red light and near-infrared light become increasingly similar during movements, so that the *measured ratio R* (5.18) is driven towards 1 and thus the *calculated oxygen saturation* converges to 84 % ((5.21), see also Fig. 5.19); compare with the similar impact of methemoglobin on the estimated saturation level from section "Specific Issues" in Sect. 5.1.2.3.

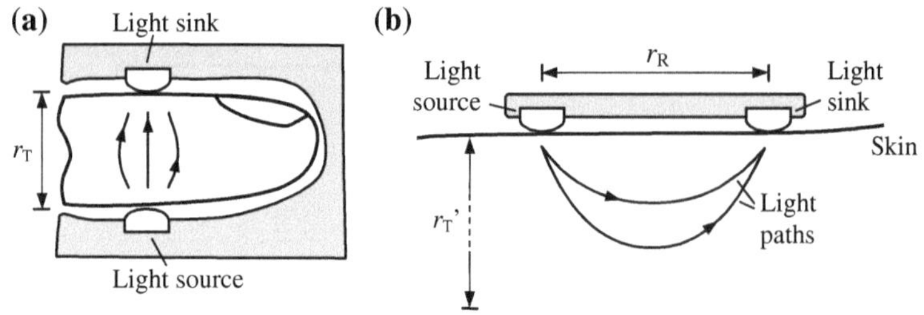

Fig. 5.22 Different arrangements of the optical sensor on the skin (Fig. 5.25). (**a**) Transmittance mode using opposite arrangement of the light source and light sink, e.g., on a finger (Fig. 5.1). (**b**) Reflectance mode using adjacent arrangement of the light source and light sink on the skin surface (Fig. 5.9a)

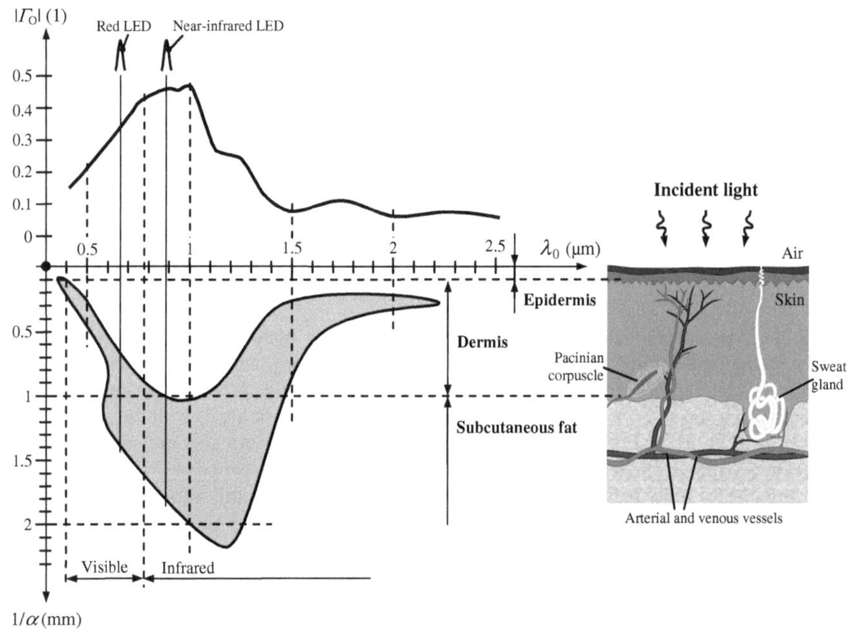

Fig. 5.23 Penetration depth $1/\alpha$ (5.10) of visible and infrared radiation into the human skin as a function of the wavelength λ_0 in free space (*left lower subfigure*). The *grey area* indicates the spread of reported depths $1/\alpha$ with the data being partially taken from ICNIRP (2006). Approximate thickness of the skin layers (compare Table 5.1) are given for comparison with the morphology of the skin (*right subfigure*). In addition, the reflection factor Γ_O (5.11) of the skin is provided (*left upper subfigure*), as adopted from ICNIRP (2006), Clark (1953), Hardy (1956). The locations and approximate intensity spectra of both wavelengths at 660 and 890 nm, i.e., red light and near-infrared light, respectively, are indicated, as emitted by light-emitting diodes (LEDs) and usually applied in the optical sensor (compare Fig. 5.8)

In addition, it can be expected that *venous blood displaced* in synchrony with motions may *overwhelm arterial pulsations* in tissue, especially if the *motion rate is similar to* f_C. Consequently, the ratio *R* would mainly consider *oxygen saturation in venous blood* but not *in arterial blood*. It means that the *calculated oxygen saturation* would tend to decrease from (relatively high) arterial saturation to (relatively low) venous saturation because of (exercise) movements; compare Fig. 3.19. Moreover, provided that tissues are *poorly perfused* and blood flow is relatively low, the *oxygenation of peripheral venous blood* can reach very low levels. In consequence, the calculated oxygen saturation may drop substantially below 84 % during movements. Likewise, *low accuracy of oximeter* or even its failure can be expected

- *during motion artefacts* and
- *poor perfusion* of transilluminated tissues (Footnote 46).

Figure 5.21a demonstrates a temporary *drop in blood oxygenation* in response to *finger movements*, whereas the corresponding optical sensor was applied on the finger subjected to movements. A few seconds after the onset of the movements, the calculated oxygen saturation starts to drop and recovers only after these movements have ceased. The resulting *time delay* in the response of the calculated saturation, i.e., its delayed drop and its delayed (subsequent) recovery, is partly due to *averaging procedures* in the oximeter; for details see section "Specific Issues" in Sect. 5.1.2.3.

Local motions and their impact on measured *optic biosignals* and on the derived *blood oxygenation* should be considered when designing *robust diagnostic monitoring*. In fact, *motion artefacts* can be *detected and reduced* using signal processing methods, e.g., as reviewed in Krishnan (2010). Artefacts can be identified and compensated by *advanced models* to estimate blood oxygenation (Goldman 2000) or even *adaptive techniques* with *reference inputs* revealing (noisy) movements only (Comtois 2007). Multiple light paths at different tissue depths, a sort of an adaptive combination of reflectance and transmittance modes[48] (Fig. 5.22), can also be utilized here in order to get rid of motion artefacts (Asada 2003; Sola 2007b).

[48] For instance, an optical sensor in the *reflectance mode* (Fig. 5.22b) can be used as a *motion sensor* (Asada 2003); see text for details. The output of such sensor can act as a motion (noise) reference for a motion (noise) cancellation filter, which is supposed to recover the true motion-free (artefact-free) optic biosignal (compare Fig. 5.21c) from a corrupted optic biosignal (Fig. 5.21b).

- It should be recalled that the *reflectance mode* is more sensitive to motions than the *transmittance mode* (section "General Issues" in Sect. 5.2.1.2).
- The *source-sink distance* of the motion sensor must be relatively short in order to get a relatively small penetration depth of light (5.26) and short light paths. It ensures that only peripheral superficial tissues are probed by light, i.e., those tissues which are easily affected by motions (see above).
- The *application site* of the motion sensor should be distant from arteries and close to veins instead.
- The employed *wavelength of light* should be selected in the red region of the spectrum such that the propagating light is highly sensitive to *motion-displaced venous blood*, i.e., sensitive to

5.2 Sensing Aspects

According to Figs. 5.1 and 5.2, *sensing aspects* of the induced optic biosignal include

- *coupling* of the transmitted light (emanating from biological tissue) into the light sink applied on the skin at a certain distance from the incident light source and
- *conversion* of the transmitted light intensity into an electric signal within the light sink.

Registered optic biosignals are the result of the incident light and its travelling through living tissues. As discussed in Sect. 5.1.2.3, the *light absorption* in tissue is modulated by (relatively *fast*) fluctuations in the local *blood volume* residing in the light propagation path (Fig. 5.1a, b) and (relatively *slow*) fluctuations in the density of dominant *chromophores* in tissue (Fig. 5.1a, c).

In fact, *three technologies* dominate in the assessment of optic biosignals, irrespective of the design of the optical sensor (Sect. 5.2.1.2):

- *spectrometry,*
- *optical plethysmography,* and
- *optical oximetry.*

The *spectrometry* utilises the basic physical principle that the absorption of light by a chromophore in tissue (Footnote 21) depends on the *density of this chromophore* (5.4) and the applied light wavelength (Fig. 5.1a, c). Likewise, the light absorption spectrum—as a function of the wavelength—provides a *signature of the chromophore type and its amount* at the sensor site to monitor the local environment of tissue.

The *optical plethysmography*, in contrast, detects variations in the light absorption in tissue (Fig. 5.1a, b). These variations mainly arise due to *pulsating volume of arterial blood* (with each heart beat) in the transilluminated region (Sect. 5.1.2.3). It should be recalled that blood is the *principle absorber in tissue* (section "Volume Effects" in Sect. 5.1.2.2). The total light absorption varies with changing *optical path lengths* through individual tissue layers (with different absorption characteristics) in living tissue (Fig. 5.14b). For instance, the optical path length through blood (vessels) in Fig. 5.1b is longer than in Fig. 5.1a, whereas the optical path length through tissue surrounding this blood vessel increases from Fig. 5.1b to a. Obviously, the *total optical path length* of the light beam in tissue *remains almost constant*; compare the accuracy and resolution aspects from section "Specific Issues" in Sect. 5.1.2.3. In fact, registered variations in the transmitted light intensity provide a *signature of (blood) volume changes* in the transilluminated region. These variations allow the registration and monitoring of *cardiac activity* (section "Cardiac Activity" in Sect. 5.1.2.3) and *respiratory activity* (section "Respiratory Activity" in Sect. 5.1.2.3).

(Footnote 48 continued)
deoxygenated blood, namely, deoxyhemoglobin; compare a relatively high μ_A applicable for red light and deoxygenated blood in Fig. 5.8.

The *optical oximetry* combines both technologies, namely, *spectrometry and optical plethysmography* to provide *blood oxygenation* in a non-invasive way, as described in section "Blood Oxygenation" in Sect. 5.1.2.3. In particular, the level of *the hemoglobin oxygen saturation* in (pulsatile) *arterial blood* is estimated, whereas hemoglobin acts as the principle chromophore in blood (section "Volume Effects" in Sect. 5.1.2.2). In accordance with the technology of the spectrometry, the light absorption in blood provides the degree of hemoglobin oxygenation (Fig. 5.8). In addition, in accordance with the optical plethysmography, the pulsatile nature of the transmitted light intensity is used in order to separate the absorption by the *pulsatile arterial blood* from the (dominating) absorption by other *non-pulsatile absorbers* (Fig. 5.14b). The registration of blood oxygenation while exploiting arterial pulsations is also known as *pulse oximetry*.[49] In other words, the pulse oximetry uses the *cardiac rhythm as a filter* and this is why the term "pulse oximetry" was coined for it.

In terms of (5.4), the absorption coefficient μ_A is the parameter of interest within the scope of the *spectrometry*, whereas the path length r is the output parameter of the *optical plethysmography*. Both the coefficient μ_A and the pulsatile change in r are relevant for the *pulse oximetry* (5.17).

5.2.1 Coupling of Light

The optical sensor is usually composed of a *light source* and a light sink which is located at a certain distance from the source, as illustrated in Fig. 5.22. The sensor is applied on the skin so that the incident light (Sect. 5.1.1) crosses the skin and enters perfused tissue below the skin. In the course of the *light propagation* in tissue, the *light interacts with biological tissue* (Sect. 5.1.2.2) and is subjected to diverse modulation phenomena (Sect. 5.1.2.3). A fraction of the modulated light leaves tissue and then enters the *light sink* which converts the intensity of the transmitted light into an *electric signal* (Sect. 5.2.1.3).

5.2.1.1 Penetration and Probing of Light

The depth to which the incident light penetrates tissue (down the skin surface) is quantitatively described by the light *penetration depth* and the light *probing depth*. These are important characteristics of the optical sensor applied within the scope of the spectrometry, plethysmography, or (pulse) oximetry (Sect. 5.2). In particular,

[49] The *pulse oximetry* solved many problems, such as low *accuracy* in the *oxygenation assessment* due to significant light absorption by tissue components other than blood (Fig. 5.14b), inherent to *oximetry* in the past and is the method used today (Wukitsch 1988); compare Footnote 41. The predecessor of the pulse oximetry did not use the pulsatile nature of the transmitted light intensity and operated in terms of the *spectrometry*.

- the *penetration depth* (section "Penetration Depth" in Sect. 5.2.1.1) is an important characteristic of the *transmittance mode* (Fig. 5.22a) in the operation of the optical sensor (Sect. 5.2.1.2). In contrast,
- the *probing depth* (section "Probing Depth" in Sect. 5.2.1.1) is a relevant characteristic of the *reflectance mode* (Fig. 5.22b).

Penetration Depth

The light *penetration depth* is defined as the tissue depth at which the incident light intensity has exponentially fallen by about 63 % (or to $1/e$), as illustrated in Fig. 5.6. The penetration depth is given by the *term* $1/\mu_A$ on the assumption that the *light scattering* in tissue can be neglected (5.4). However, as discussed in section "Inhomogeneity Effects" in Sect. 5.1.2.2, the light scattering is by far the dominant *tissue-photon interaction* in biological media in comparison with the light absorption. In the latter case, the *photon diffusion theory* yields the *term* $1/\alpha$ *as the penetration depth* (5.8) which is then a function of both μ_A and μ_S' (5.10).

Figure 5.23 demonstrates *reported penetration depths* over the wavelength λ of *visible and infrared radiation* incident on the *human skin*. In particular, the depth increases in the visible range with increasing λ and then strongly decreases in the infrared range. It can be expected that the light absorption by *blood and melanin* determines the latter behaviour of the penetration depth in the visible range (section "Volume Effects" in Sect. 5.1.2.2). In the infrared range, increasing (on average) light absorption by *water* with increasing λ determines the skin absorption; compare Fig. 5.7. Consequently, the relatively long infrared wavelengths are only absorbed by the outermost layer of the skin, the *epidermis* (Fig. 5.23). As a practical approximation,

- the *penetration depth* of *several millimeters* results for λ in the range of 780–1,400 nm,
- *less than 1 mm* for 1,400–3,000 nm (see Fig. 5.23), and
- *less than 0.1 mm* and even down to a few micrometers for 3 μm−1 mm (ICNIRP 2006).

Figure 5.23 also reveals that minima and maxima of the *penetration depth* $1/\alpha$ almost coincide with the reflection minima and maxima, respectively. Likewise, the *reflection minima and maxima* coincide with *maxima and minima of the total attenuation* α; for details see section "Inhomogeneity Effects" in Sect. 5.1.2.2.

Probing Depth

The light *probing depth* is defined as the probed tissue depth at which (pulsatile) changes in local optical properties yield (pulsatile) changes of at least 5 % of the (absolute) transmitted light intensity, as registered by the light sink (Fawzi 2003).

The *mean probing depth d* or the mean depth of the transilluminated *banana-shaped region*[50] *of the penetrated tissue*—as illustrated in Figs. 5.9b and 5.20—can be estimated using the following *numerical equation* (Mannheimer 1997; Franceschini 1999a), to give

$$d = \frac{0.4 \cdot \sqrt[2]{r_R}}{\sqrt[4]{\mu_A \cdot \mu_S'}}. \tag{5.26}$$

Here the physical dimension of the depth d and the source-sink distance r_R is 1 mm, whereas the dimension of the coefficients μ_A and μ_S' is 1/mm; compare Table 5.1. The depth d evidently increases with the *separation distance* r_R, as also demonstrated in Fig. 5.20 with $d' > d$ and $r_R' > r_R$. On the other hand, the depth d decreases with increasing light absorption and scattering strength in tissue. Likewise, *identical probing* (depth) of two light colours results when the corresponding wavelengths yield *identical products* $\mu_A \cdot \mu_S'$; this is an important aspect of the accuracy and resolution of the oximetry (see section "Specific Issues" in Sect. 5.1.2.3). In analogy to (5.26), the *maximum depth* of the transilluminated banana-shaped region from the skin surface can be approximated as $r_R/2$ (Wang 2012).

From a *physiological point of view*, the depth d is strongly influenced by *blood oxygenation* and the local *blood perfusion* as well as the dominance of the *adipose tissue* layer *under the skin*. In fact, these physiological characteristics determine the effective size of μ_A, μ_S' and thus the size of the denominator in (5.26), as discussed below.

Figure 5.17 illustrates the change in the mean *probing depth* in perfused tissue with

- varying *blood perfusion* in the periphery (over the respiration cycle),
- varying *colour* of the incident light (from red to near-infrared light), and
- varying (hemoglobin) *oxygen saturation* in arterial blood (from $S = 0$ to 100 %).

As discussed in section "Specific Issues" in Sect. 5.1.2.3 (see accuracy and resolution aspects), increasing local *blood perfusion* tends to decrease the probing depth and the propagation distance of light in tissue. This is because *blood is the principle absorber* in perfused tissue (Footnote 20) and thus the effective light absorption by tissue increases with the volume fraction of blood in tissue. Consequently, the higher is the effective light absorption, the less is the probing depth; compare decreasing d with increasing μ_A in (5.26).

[50] In fact, the *banana-shaped region* of the penetrated tissue delimits the pathway of photons which *detection probability* is the highest at the *light sink* (pathway A in Fig. 5.9b). Consequently, other pathways show lower detection probabilities (pathway B in Fig. 5.9b). In other words, if a propagating photon remains within this banana-shaped region during its *random walk* through tissue, it will be detected by the sink. However, if a photon is scattered out of this banana-shaped region (pathway C in Fig. 5.9b), it is likely that this photon will no longer be accessible for the sink.

It can be observed from Fig. 5.17 that the absolute *probing depth* and its *relative increase during inspiration* (due to *decreased blood volume* in the periphery, section "Respiratory Activity" in Sect. 5.1.2.3) are *inversely related* to each other. In analogy to the discussion about the pulsatile deflection amplitude in Fig. 5.19 (section "Blood Oxygenation" in Sect. 5.1.2.3), the higher is the effective μ_A of blood (for given values of S and λ, Fig. 5.8), the less is the probing depth, and the stronger is the corresponding respiration-induced decrease in the (local) *light absorbance from the expiration to inspiration phase.* Consequently, the stronger is the respiration-induced decrease in the light absorbance *from expiration to inspiration*, the greater is the associated relative increase of the *probing depth*. Obviously, this observation is valid only for *homogenous and perfused tissue* under the skin surface.

In particular, blood perfusion of the *upper dermal layers* has a significant effect on the *probing depth* (Meglinski 2002) since the optical sensor in the *reflectance mode* is usually applied on the skin surface. With increasing perfusion strength of the upper dermal layers, the light intensity in the deep tissue layers decreases (situated below the dermal layer), as well as the effective probing depth.

The effects of *light colour* and *oxygen saturation* on the *probing depth*—as shown in Fig. 5.17—can be easily interpreted by considering the impacts of the colour and saturation on the effective size of μ_A (Fig. 5.8). In short, the higher is μ_A of blood (for given values of S and λ, Fig. 5.8), the stronger is the light absorption in perfused tissue, and thus the shallower is the resulting probing of light.

Figure 5.20 illustrates the behaviour of the *light penetration paths* and *probing depths* in perfused tissue from another perspective. According to the theoretical modelling of the *reflectance mode* using the photon diffusion theory (Mannheimer 1997), the mean depths d for *red light* ($\lambda = 660$ nm) and *near-infrared light* (890 nm) are almost equal for *high* $S = 100$ % and amount to about 4 mm. In this case, the *probability distributions* (histograms) of the probing depths (of individual light beams) are very similar for red and near-infrared light, extend up to the depth of 10 mm, and are relatively wide. The corresponding behaviour of the light penetration paths is shown by solid lines in Fig. 5.20. In contrast, the widths of such probability distributions for *red and near-infrared light* strongly diverge for *low* S. In particular, the distribution width for red light becomes smaller than that for near-infrared light, whereas the propagating red light (in tissue) becomes confined to the skin surface. The mean depth d becomes 3 and 4 mm for red and near-infrared light, respectively, at low $S = 40$ %; compare dashed lines in Fig. 5.20. As already discussed, this is because μ_A of deoxyhemoglobin (low S) for red light is much larger than for near-infrared light, and, on the other hand, μ_A of oxyhemoglobin (high S) changes less strongly from red light to near-infrared light (Fig. 5.8).

Heterogeneous tissue composition plays an important role in the light probing. For instance, the relevance of the superficial (subcutaneous) *adipose tissue* layer—typically residing *under the skin surface* and above the *muscular layer*—in terms of the *probing depth* is shown in Franceschini (1999a).

- The authors demonstrate that the *estimated effective values of* μ_A—estimated out of the incident and transmitted light intensities in the reflectance mode—are relatively high for *low thickness of the adipose tissue* layer of only 3 mm and differ only *slightly* with *varying distance* r_R (i.e., μ_A increases slightly with increasing r_R). It is suggested that the underlying skeletal *muscle* is predominantly probed by the incident light even at the shortest r_R employed (r_R = 10 mm); compare Fig. 5.24a.
- In contrast, the estimated effective values of μ_A are much lower for *high thickness of the adipose tissue* of about 10 mm (> 3 mm), which indicates the light absorption by the probed adipose layer. Here it should be recalled that *subcutaneous fat* exhibits μ_A which is only about one third of μ_A of *muscle* (with a relatively high fraction of blood, the strongest absorber); see Table 5.1. In the latter case of the thick adipose layer, the estimated μ_A also increases only *slightly* with increasing r_R. It suggests that the (superficial) adipose layer is mostly probed even at the largest r_R employed (r_R = 30 mm); in view of the tendency of the probing depth to increase with increasing r_R (compare Figs. 5.24c and (5.26)).
- In the case of an *intermediate thickness of the adipose tissue layer* of about 6 mm (> 3 and < 10 mm), the estimated μ_A and the corresponding probing depth indicate a predominant probing of the adipose layer for relatively *short* r_R (i.e., estimated *low* μ_A and *shallow probing*) and a predominant probing of the underlying muscle layer for relatively *large* r_R (i.e., estimated *high* μ_A and *deep probing*); compare Fig. 5.24b. Likewise, a *shallower probing* (depth) is usually related to a *lower* μ_A affecting the probing light.

The *light scattering* impacts also the formation of the light penetration paths and the resulting *probing depth*. Experimental data in (Franceschini 1999a) indicate that the *estimated effective values of* μ_S' tend to be larger for the (subcutaneous) *adipose tissue* than the *muscular tissue*, especially for large distances r_R > 27 mm applied. In fact, it confirms the quantitative scattering data from Table 5.1. Furthermore, the estimated values of μ_S' are subject to a stronger change over the varying r_R for a

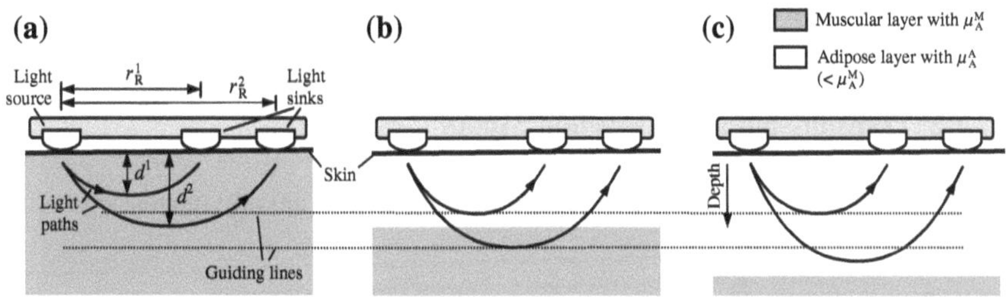

Fig. 5.24 Mean probing depth d as a function of the source-sink distance r_R in the reflectance mode (Fig. 5.22), provided a two-layered tissue below the skin. (**a**) Muscular tissue directly under the skin with the absorption coefficient μ_A^M. (**b**) Relatively thin layer of the subcutaneous adipose tissue with μ_A^A (< μ_A^M, Table 5.1), which resides between the skin surface and muscular tissue. (**c**) Relatively thick layer of the subcutaneous adipose tissue

relatively *thick layer* of the adipose tissue in comparison with a relatively *thin adipose layer*. Likewise, the scattering is pronounced and its strength is sensitive to the applied r_R when the propagating light probes predominantly the relatively thick adipose layer. In contrast, the predominant probing of the *muscular layer*—which is located under the (relatively thin) *adipose layer*, Fig. 5.24b—exhibits less light scattering and its sensitivity to varying r_R is less. Furthermore, the *spectral dependence* of μ_S' (i.e., μ_S' increase with decreasing λ, section "Inhomogeneity Effects" in Sect. 5.1.2.2) is less pronounced at shallower probing depth (with the adipose layer probed) than deeper in tissue (the muscular layer probed), as shown in Franceschini (1999a).

Figure 5.24 illustrates the discussed effects of the *subcutaneous adipose tissue* on the *probing depth* of the incident light. In the case of the *dominant muscular tissue under the skin* (Fig. 5.24a), the resulting probing depths d^1 and d^2 (> d^1) for the distances r_R^1 and r_R^2 (> r_R^1), respectively, are relatively low in comparison with depths within the *dominant adipose tissue under the skin* (Fig. 5.24c). This is because the depth d is inversely related to the product $\mu_A \cdot \mu_S'$ (5.26) which is about 0.02 and 0.01 mm^{-2} for the muscular tissue and adipose tissue, respectively (as estimated from Table 5.1). Likewise, the strong light absorption by the well-perfused muscular tissue determines shallow depth d. Figure 5.24b also shows the case in which the muscular tissue is only probed at the large distance r_R^2.

From a practical point of view, a thick layer of the subcutaneous *adipose tissue* limits disadvantageously the *probing depth* of the incident light beam to the spatial extent of the adipose layer (Fig. 5.24c) and increases the effective *scattering* of light in tissue. That is, the thick adipose layer may hinder the probing of (deeper located) *perfused tissues* such as the *muscular tissue* (Fig. 5.24b, c). For instance, it may become a key aspect if the optical sensor is applied on the *chest skin* where a thick layer of the subcutaneous adipose tissue can be expected (Kaniusas 2006, 2007); compare Footnote 55.

Finally, another practical issue should be shortly mentioned. As shown in (5.26), an increase in r_R tends to increase the mean depth d; however, the *transmitted light intensity* at the *light sink* progressively decreases too. This means that the aimed maximum value of d, i.e., the aimed *probing of deep tissues* under the skin surface, while maximizing the distance r_R, is restricted by the acceptable *signal-to-noise ratio* of the transmitted light intensity (provided a constant power supply of the light source).

5.2.1.2 Transmission and Reflection Modes

The *optical sensor on the skin* can be operated in the two already mentioned modes (Fig. 5.22):

- *transmittance mode* and
- *reflectance mode*.

These modes basically differ in the way the light photons are collected by the light sink, as discussed in the following chapters.

Table 5.3 Qualitative comparison of the transmittance and reflectance modes concerning the application of the optical sensor on the skin, see Fig. 5.22

Mode of operation	Application region	Pulsatile fraction $\mathcal{R}$	Signal-to-noise ratio	Sensitivity to motion artefacts	Registered photon flux
Transmittance mode	Distal	High	High	Low	Forward
Reflectance mode	Distal and proximal	Low	Low	High	Forward and backward

General Issues

As illustrated in Fig. 5.22, photons can be collected either with a light sink on an *opposite skin surface* (Fig. 5.22a) to the surface of the incident light or on an *adjacent skin surface* (Fig. 5.22b). The former case is known as the *transmittance mode* which utilizes the *straightforward light path* from the light source through tissue to the light sink. The latter case is known as the *reflectance mode* in which the light sink receives reflected, more accurately, reemitted light; i.e., there is no straightforward light path. Likewise,

- the *sink* in the *transmittance mode* receives light *aligned in the direction* of the incident light, whereas

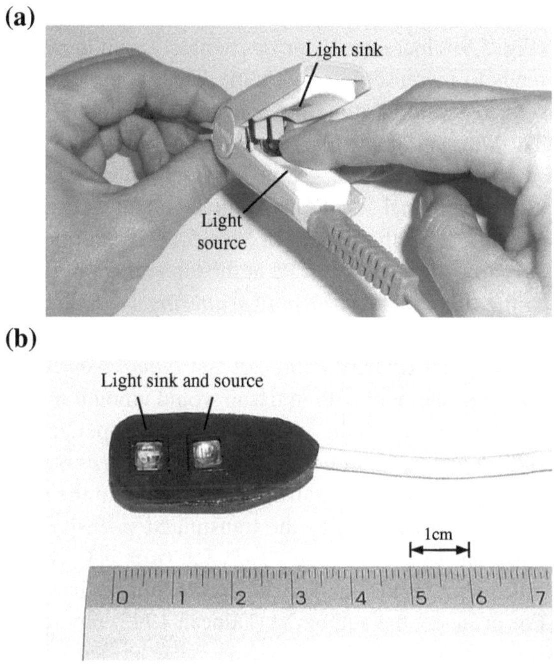

Fig. 5.25 Practical realisations of optical sensors operated in the different modes; compare Fig. 5.22. (**a**) Transmittance mode. (**b**) Reflectance mode

- the *sink* in the *reflectance mode* receives light *aligned in the opposite direction* of the incident light (Table 5.3);

compare light paths in Fig. 5.22a, b. In an approximation, the transmittance mode shows a *relatively deep penetration* of light into tissue, whereas the reflectance mode shows a *relatively shallow probing* of light (Sect. 5.2.1.1).

Figure 5.25 illustrates practical designs of *optical sensors*. The shown sensor in the transmittance mode is realised as a *finger-mounted clip sensor* (Fig. 5.25a), whereas the sensor in the reflectance mode is realised as a *skin-mounted plaster-like sensor* (Fig. 5.25b).

The incident light *interacts with tissue* (Sect. 5.1.2.2) in the course of which the light intensity undergoes diverse *physiological modulations* (Sect. 5.1.2.3). In fact, the *transmitted light intensity*—as measured by the light sink—depends on the composition of biological tissues, especially the volume fraction of *blood* and its *oxygenation state* (see the technology of spectrometry, Sect. 5.2), and the *path length of light* through individual tissue layers (see the technology of optical plethysmography, Sect. 5.2).

It can be expected from (5.4), that the *transmitted light intensity* decreases nearly exponentially as the *source-sink separation distance* r_T or r_R (Fig. 5.22) increases. This approximation is rather applicable for the *transmittance mode* with the straightforward light path. However, the authors in (Mendelson 1988) prove also nearly exponential decrease of the transmitted light intensity in the *reflectance mode* as r_R increases; in particular, exponential decrease of both *pulsatile and non-pulsatile components* as r_R increases.

In the *reflectance mode*, the effective length of the *banana-shaped light path* (Fig. 5.9b) increases disproportionately with increasing r_R because the probing depth tends to increase with r_R (5.26). That is, the effective light paths are longer for the reflectance mode than the transmittance mode, assuming equal distances $r_R = r_T$. The *photon diffusion theory* (Schmitt 1991) indirectly confirms the latter conclusion in showing that calibration curves for the assessment of blood oxygenation are almost identical for the transmittance and reflectance modes provided that $r_T > r_R$ (section "Specific Issues" in Sect. 5.1.2.3). Furthermore, it should be noted that only very few photons penetrate at the top surface of the skin in the reflectance mode due to the dominant *anisotropic scattering* in biological tissues (compare large $g \gg 0$ from Table 5.1), as discussed in section "Inhomogeneity Effects" in Sect. 5.1.2.2.

In a first *approximation* of the *banana-shaped light path as a semicircle*, the length of the light path in tissue would amount to $\pi \cdot r_R$ ($\approx 3.14 \cdot r_R$). Consequently, the transmitted light intensity would experience a stronger attenuation in the *reflectance mode* in comparison with the *transmittance mode* with equal *distances* $r_R = r_T$. Namely, the transmitted intensity in the reflectance mode is proportional to $e^{-\pi \cdot \mu_A \cdot r_R}$ (5.4), whereas the transmitted intensity in the transmittance mode is proportional to $e^{-\mu_A \cdot r_T}$ ($> e^{-\pi \cdot \mu_A \cdot r_R}$ for $r_R = r_T$). The reality is that light paths in the reflectance mode tend to be much longer than r_R and even longer than $\pi \cdot r_R$. For instance, the authors in (Duncan 1995) report that light paths are even 4–6 ($> \pi$)

times longer than r_R in the human head and arm at $\lambda = 807$ nm and $r_R \approx 4$ cm; compare with the *differential pathlength factor* from section "Inhomogeneity Effects" in Sect. 5.1.2.2.

In general, the principles of the *transmittance and reflectance modes* are similar. However, there are some important differences from an application point of view, as summarized in Table 5.3. In contrast to the transmittance mode, the reflectance mode yields a poorer *signal-to-noise ratio*, mainly because of a relatively shallow *light probing*. The missing straightforward light path in the reflectance mode implies that the spatial course of the *light path* is less defined and this course may strongly differ for red and near-infrared light (Fig. 5.20). This is because of *multiple scattering* and *heterogeneous tissues* under the skin surface; compare sections "Specific Issues" in Sect. 5.1.2.3 and "Probing Depth" in Sect. 5.2.1.1. Furthermore, the reflectance mode is more susceptible to *(motion) disturbances* than the transmittance mode (Table 5.3), see below. It should be stressed that the *light scattering* is a *favourable phenomenon for the reflectance mode* since there is no direct light path from the light source to the sink, whereas the scattering is *unfavourable for the transmittance mode* with a direct light path (Fig. 5.22).

The *pulsatile fraction* $\mathcal{R}$ of the transmitted light intensity due to the pulsatile arterial blood (5.17) is a good indicator of the quality of the optic biosignal. The fraction $\mathcal{R}$ is also known as the *blood perfusion index*, estimated as the ratio of the pulsatile to average signal level. This fraction $\mathcal{R}$ amounts to only *a few percent* in the *transmittance mode* (compare Fig. 5.15c); in particular, this fraction resides in the range of 0.02–0.05 (König 1998). In contrast, the *reflectance mode* yields much lower values of $\mathcal{R}$ in the range of 0.001–0.005 (Table 5.3). That is, the fraction $\mathcal{R}$ in the reflectance mode is about 10 times weaker than in the transmittance mode.[51]

Furthermore, the *pulsatile fraction* $\mathcal{R}$ in the *reflectance mode* increases with *increasing* r_R (Ling 1993) because the *probing depth* increases and the light beam progressively probes deeper and better perfused tissue layers (Fig. 5.24b). An almost *linear relationship* between $\mathcal{R}$ and r_R (with r_R in the range of 4–11 mm) for red and near-infrared light was reported in Mendelson (1988). Here the incident light intensities were adjusted (increased) such that for each (increased) distance r_R the non-pulsatile component of each transmitted light intensity remained constant. Obviously, the *signal-to-noise ratio* tends to decrease with *increasing* r_R (and thus with increasing both *light path* and the *total light absorption* in tissue, 5.4) provided a constant *power* of the incident light. There is a *tradeoff* between the (large) distance r_R proportional to (advantageously large) $\mathcal{R}$ and, on the other hand, the (advantageously low) power supply of the light source. The *low power consumption* of the light source is necessary to avoid *skin burning problems* during

[51] It should be noted that *near-infrared light* would be more appropriate to assess the *pulsatile fraction* $\mathcal{R}$ (or the *blood perfusion index*) as compared with red light. This is because the light absorption changes minimally for near-infrared light within the normal range of blood oxygenation, as compared with red light; see Fig. 5.8.

long-term applications of the optical sensor (Asada 2003) and to allow wearable long-term monitoring; compare Sect. 5.2.1.4. It follows from the above that the optimal value of r_R can be expected, which yields sufficient levels of both the fraction $\mathcal{R}$ and the signal-to-noise ratio, see section "Source-Sink Distance" in Sect. 5.2.1.2.

In accordance with (5.17), the *fraction* $\mathcal{R}$ is proportional to the pulsatile deflection Δr_p of arterial blood, namely, to the systolic-diastolic deflection of *arterial vessels* (Fig. 5.14c). Likewise, this fraction is proportional to the pulsatile deflection of *capillary beds* with each pressure pulse, i.e., to a consecutive increase and decrease of the total diameter of the transilluminated capillary beds. This interpretation of $\mathcal{R}$ is particularly important when considering the *reflectance mode* with its predominant probing of peripheral superficial capillary beds. High values of $\mathcal{R}$ indicate good *blood perfusion* and *high density of capillaries* in the region probed by light.

In addition, *reflections* of the incident light, e.g., on the *entry of light into the skin* or cranial bone, may play a significant role (König 1998); e.g., with the optical sensor in the *reflectance mode* applied on the forehead. A (small) fraction of the *direct light* which propagates from the light source to the light sink in the air above and close to the skin surface may also impair the proper operation of the *reflectance mode* (König 1998; Asada 2003). Consequently, *air gaps* between the skin surface and optical components of the sensor should be avoided. In addition, (very) small distance r_R favours this fraction of the unwanted direct light which may even saturate the light sink.

To avoid this *direct light* in the *reflectance mode*, the incident light beam should be *focussed* normally to the skin surface by the *light source*; in analogy, the *light sink* should show the highest sensitivity to the incoming light normal to the skin surface. However, strong *directional properties* of the source and sink show *adverse effects* when the optical sensor experiences a movement relative to the skin, i.e., a *motion artefact*, as described in section "Motion Artefacts" in Sect. 5.1.2.3. The movement could deflect the direction of the incident light as well as the sensing direction of the sink leading to strong fluctuations in the optic biosignal in view of *heterogeneous tissue* structures under the skin surface (Fig. 5.9a).

As already noted in section "Specific Issues" in Sect. 5.1.2.3, the *reflectance mode* can be more easily compromised by *poor perfusion* (e.g., by *vasoconstriction*) than the transmittance mode. This low robustness of the reflectance mode could be explained by its *superficial light probing* combined with (particularly) strong decrease of the local peripheral perfusion in the superficial tissue layers during vasoconstriction. In other words, the propagating light in the *transmittance mode* penetrates to *greater depths*, larger blood vessels, and better perfused tissues—which in fact are less affected by vasoconstriction—in comparison with the reflectance mode.

In addition, *motion artefacts* or *external pressures* affect easily peripheral *superficial capillary beds* (section "Motion Artefacts" in Sect. 5.1.2.3). Their (partial) *collapse* with increasing external pressure yields a significant change in the measured optic biosignal with the optical sensor operated in the *reflectance mode*

(Asada 2003). In contrast, the measured biosignal from the *transmittance mode*—with the relatively *deep penetration of light* into tissue—does not significantly change when some superficial capillary beds are collapsed (Table 5.3).

Source-Sink Distance

Concerning the operation of the optical sensor in the *reflectance mode*, the optimal selection of the *source-sink distance* r_R (Fig. 5.22b) is highly relevant and thus will be discussed here. As already noted, the *probing depth* tends to increase with *increasing* r_R (5.26) so that deeper and better *perfused tissues* are progressively probed; compare Fig. 5.24 and section "Probing Depth" in Sect. 5.2.1.1. In addition, a large probing depth is desired in order to attain high $\mathcal{R}$ (5.17). For instance, the authors in (Ling 1993) confirm an increase in $\mathcal{R}$ for red light (660 nm) and near-infrared light (830 nm) as the distance r_R increases; although the absolute *intensity of the transmitted light* decreases with r_R as well. However, the *person to person variability* of the light paths in tissue increases with increasing both r_R and the probing depth, because the *heterogeneity of tissues* under the skin surface becomes increasingly important (Takatani 1994); compare Fig. 5.24.

From a quantitative point of view, the distance r_R *can not be too small* (< 1 mm) because then the transmitted light intensity would carry little information about the absorption and scattering properties of tissues along the photons' paths. The *path length of light* in tissue would be too short for *absorption* and *scattering* events to occur (in a necessarily large number) before exiting light photons are collected by the *light sink*. This is due to a (relatively long) *average free path* of a *light photon* in tissue (Winey 2006) in the range up to a few centimetres before the photon encounters the *next absorption event* (5.4) and, on the other hand, in the range of sub-millimeters before the photon encounters the *next scattering event* (5.5). It should be recalled that this free path, i.e., *average free path* length, is given by

- $1/\mu_A$ for the light absorption only (section "Volume Effects" in Sect. 5.1.2.2), by
- $1/\mu_S$ for the light scattering only (section "Inhomogeneity Effects" in Sect. 5.1.2.2), and by
- $1/\alpha$ for the light absorption and predominant light scattering (5.10).

For instance, Table 5.1 yields $1/\alpha = 5.8$ mm for (average) tissue and $\lambda = 800$ nm, i.e., for the isosbestic point in Fig. 5.8.

Furthermore, *small distance* r_R (or small distance r_T, Fig. 5.22a) favours *small power consumption* of the optical sensor, which is highly relevant in terms of a wearable biomedical sensor. In other words, an acceptable level of the *signal-to-noise ratio* requires increasing power of the light source with increasing r_R (section "General Issues" in Sect. 5.2.1.2).

Optimal values of r_R were shown to be in the range 2–5 mm for biological tissue in the *near-infrared* band (800–2,500 nm), characterized by

- a weak dependence of the transmitted light on the *light scattering variations*[52] in tissue (with the minimum sensitivity at around 2.5–3 mm) and
- a strong dependence of the transmitted light on the *light absorbance variations* in tissue (Kumar 1997).

These optimal values of r_R decrease with increasing absolute values of μ_S' in tissue. In particular, the *reflectance* (i.e., *photon flux* exiting tissue in the region of the *photon's sink*) increases as μ_S' increases for small source-sink separation ($r_R < 1/\alpha$) because of dominating *backward-propagating flux*. Likewise, the probability of *backscattering increases with* μ_S'. For separation distances larger than about 2.5–3 mm, the *reflectance begins to decrease with increasing* μ_S' because *forward-propagating flux* starts to dominate. The reflectance measured at large values of r_R ($r_R > 1/\alpha$) is equally sensitive to changes in both scattering and absorption properties of the transilluminated tissue.

In other words, the *light sink captures* predominantly the *backward-propagating flux* of photons at relatively small r_R and the *forward-propagating flux* at relatively large r_R (Kumar 1997); compare Table 5.3. Here the optimal distance r_R between the light source and sink is set in a way that balances the backward-propagating flux and the forward-propagating flux of the re-emitted photon fluence. Concerning the highest *signal-to-noise ratio* and the *robustness* of *R* (5.18) against *artefacts*, the optical sensor with $r_R = 7$ mm proved to be favourable (Ling 1993).

Multiple light sources and sinks are suggested to increase the level of the *transmitted light intensity* in the *reflectance mode* as well as to minimize the effects of the *tissue heterogeneity* (Takatani 1994); see section "Probing Depth" in Sect. 5.2.1.1. To give a few examples, a ring-shaped photodiode embracing a single LED was realized in Duun (2007), multiple discrete photodiodes were incorporated around a single LED to enhance the transmitted signal level (Takatani 1994; Mendelson 1988), or even multiple LEDs were arranged around a single photodiode to attain data on local light paths (Sola 2007a).

Light Wavelength

As discussed in section "Volume Effects" in Sect. 5.1.2.2, *optical sensors* usually employ wavelengths in the *optical window*, in which biological tissues are *reasonable transparent* to the incident light. As illustrated in Fig. 5.7, such window

---

[52] In the case that *high sensitivity of the transmitted light to scattering variations* in tissue is required—variations in μ_S' may also have diagnostic value—the authors in (Kumar 1997) suggest optimal distances r_R for the *reflectance mode* (Fig. 5.22b). Namely,

- if both μ_A *and* μ_S' of tissue *are small*, the latter sensitivity has its maximum value when the distance r_R is small; i.e., the *registered reflectance increases as* μ_S' *increases*, see text. In contrast,
- if both μ_A *and* μ_S' *are large*, this sensitivity has its maximum value when the distance r_R is large; i.e., the *registered reflectance decreases as* μ_S' *increases*, see text.

resides in the range of approximately 600–1,300 nm. Obviously, the resulting transmitted light intensity is a non-linear function of not only wavelength, but also blood oxygenation, and the type of the transilluminated tissue (Table 5.1).

Typically, two wavelengths are applied, the *red wavelength* at around 660 nm and the *near-infrared wavelength* at around 890 nm (Fig. 5.8). In particular, these two wavelengths are required to estimate *blood oxygenation* out of the transmitted light intensities (5.18 and 5.21) because absorption characteristics of oxygenated and deoxygenated blood significantly differ at these wavelengths (section "Blood Oxygenation" in Sect. 5.1.2.3). In short, the ratio R of the *pulsatile fraction* of the transmitted light intensity at the red wavelength to the *pulsatile fraction* at the near-infrared wavelength is inversely related to the oxygenation level (5.18 and 5.21).

The *number of discrete wavelengths* that are used must be equal (at least) to the number of discrete absorbers in tissue (Wukitsch 1988). Thus *two wavelengths* correspond to the two *strongest absorbers* in perfused tissue: *oxyhemoglobin* and *deoxyhemoglobin* in blood (section "Volume Effects" in Sect. 5.1.2.2). In other words, an optical sensor with more than two wavelengths would principally allow detection and quantification of other (usually minor) absorbers in blood such as methemoglobin (section "Specific Issues" in Sect. 5.1.2.3 and Footnote 220 in Sect. 3). There are optical approaches which apply up to 15 different wavelengths (Franceschini 1999b; Manzke 1996).

A study by (Schmitt 1991) suggests the range of 630–670 nm for red light and 900–950 nm for near-infrared light based on the *photon diffusion theory*. Here the difference between μ_A of oxyhemoglobin and μ_A of deoxyhemoglobin (Fig. 5.8) is maximised for both red and near-infrared light, which yields *maximum resolution of the oxygenation level* (section "Specific Issues" in Sect. 5.1.2.3).

If we consider different *light colours* in relation to *optimal* r_R, the *average free path* $1/\alpha$ should be considered, as discussed in section "Source-Sink Distance" in Sect. 5.2.1.2. That is, the distance r_R should be at least in the range of $1/\alpha$, otherwise, absorption and scattering properties of tissue will not be reflected by the transmitted light intensity. In particular, the *absorption* strength in tissue determines α (5.10) and, on the other hand, depends strongly on blood oxygenation and the light wavelength applied; compare Figs. 5.8 and 5.20. The light *scattering*—which also determines α—scales inversely with wavelength. Thus photons of *red light* scatter more strongly and undergo more scattering events than photons of *near-infrared light*, before red and near-infrared photons arrive at the light sink. Photons of *red light* gather more information about absorption and scattering properties of tissue because these photons show increased effective path length; compare with the relevance of *multiple scattering* and the *differential pathlength factor* from section "Inhomogeneity Effects" in Sect. 5.1.2.2. It can also be expected that red photons dominate in the *backward-propagating flux* of photons (section "Source-Sink Distance" in Sect. 5.2.1.2).

Application Regions

There are significant differences in possible *application regions* of the optical sensor operated in the transmittance or reflectance mode. The *transmittance mode* is applicable only on a few *distal regions* such as the finger, earlobe,[53] nose, or in infants the foot and palm (Fig. 5.22a). The variety of these regions is limited by the opposite arrangement of the light source and light sink over a relatively short distance r_T; i.e., only relatively *thin parts of the body* comprise potential application regions. The upper bound of r_T is limited by the finite *penetration depth* of light (Fig. 5.6). The most typical application site is the finger (or the fingertip or phalanx); compare Footnote 37.

In contrast, the *reflectance mode* is applicable on various *proximal and distal regions* of the skin, typically on the forehead; compare Table 5.3. A planar arrangement of the light source and light sink over a relatively short distance r_R (Fig. 5.22b) favours versatile applications of the reflectance mode and the use of the

---

[53] Significant differences can be observed in *optic biosignals* recorded from the *finger* or *earlobe*, as typical application regions of the optical sensor (Fig. 5.22); compare Footnote 55. The reason for this is that the fingertip includes a large number of *arterio-venous anastomoses* (for the thermoregulatory control, see Sect. 3.1.5), whereas such anastomoses lack in the ear (Middleton 2011); see Footnote 37. Since arterio-venous anastomoses are *sympathetically innervated* (controlled), i.e., the sympathetic activation closes (blood) shunts between arterioles and venules (Fig. 3.22b), the *transmitted light intensity* from the *finger sensor* can be expected to *reflect the sympathetic vascular tone* but not from the *earlobe sensor*. In particular, pulsatile changes of the transmitted intensity that arise in the course of *pulsatile changes of the (local) blood volume* reflect this sympathetic tone. Namely, with increasing sympathetic tone in the finger, the *distensibility and radius* of anastomoses decrease. Consequently, the *local pulsatile blood volume* decreases as well as the deflection amplitude of the *pulsatile changes in the transmitted intensity* (comparable to the deflection $s_{S,D}$ in Fig. 5.15a, Sect. 5.1.2.3). Obviously, the pulsatile blood volume is strongly affected by the *distensibility of the arterial vessel*; in general, the *pulsatile volume is a product of the vessel's compliance and the pulsatile pressure* within the vessel (see 5.14, 2.23, and Sect. 2.5.1). For instance, the deflection amplitude decreases by about 48 and 2 % with the optical sensor applied on the finger and ear, respectively, in response to the vasoconstrictive *cold stimulus* (Awad 2001); compare Fig. 5.28.

The *ear* is relatively *immune to vasoconstrictive effects of the sympathetic system*, thereby minimizing the effects of local peripheral vasoconstriction on the measurements of blood oxygenation with the optical sensor on the ear (section "Specific Issues" in Sect. 5.1.2.3). The pulsatile blood volume in the ear is

- less affected by the vessel's distensibility and
- mainly responds to *pulsatile changes of the (central) blood pressure* (according to 2.23) and to changes in the systemic circulation (Awad 2001).

In addition, the optic biosignal from the *earlobe* was reported to be rather insensitive to changes in the *contacting force* (i.e., the contacting force between the optical sensor and its application site on the skin) due to movements (Middleton 2011); compare section "Contacting Force and Skin Temperature" in Sect. 5.2.1.2. That is, *motion artefacts* can be expected to be less dominant in the optic biosignal from the earlobe in comparison with that from the finger (section "Motion Artefacts" in Sect. 5.1.2.3).

optical sensor as a wearable biomedical sensor. The distance r_R determines the light *probing depth* (5.26) and the *signal-to-noise ratio* (section "General Issues" in Sect. 5.2.1.2).

In fact, only those *skin regions* are suitable as application regions of the optical sensor, which show dense and *rich capillary beds* under the skin surface (high vascularization level). The suitable skin regions must be well-perfused[54]; compare Footnote 46. However, *individual skin compositions* lead to the varying (effective) light absorption as registered by the optical sensor on the skin.

The application region of the optical sensor has a strong impact on the *response time* of the sensor to *oxygenation changes* in the lungs, as discussed in section "Specific Issues" in Sect. 5.1.2.3. The more *proximally* (centrally) the sensor is applied, the smaller is the *time delay* (i.e., response time) in the registered (estimated) blood oxygenation. In addition, this time delay increases strongly in *distal application regions* with poor blood perfusion (in peripheral regions with decreased blood flow), peripheral vasoconstriction, or hypothermia.

Besides the time delay, the *accuracy* of the *optical sensor applied distally* can significantly deteriorate during, for instance, stress-induced *centralization of blood*. In the course of the centralization, the perfusion of vital organs such as the heart and brain is maintained only. Such centralization of blood can only occur at the expense of strongly *reduced blood perfusion in the periphery* (in peripheral extremities), i.e., at the distal site where the optical sensor is applied on the skin.

The range of possible *non-standard applications* (and non-standard application regions) of the *optical sensor* is huge. In particular, applications include a sensor attached to the head of the fetus, a thoracic[55] sensor (Kaniusas 2006, 2007), a maternal abdominal sensor for the trans-abdominal monitoring of fetal blood oxygenation (Zourabian 2000), an implantable sensor (Reichelt 2008), an esophageal sensor (Kyriacou 2002), a ring sensor (Asada 2003; Sola 2006), a miniaturized sensor worn in the ear channel (Venema 2012), and an autonomous sensor powered by body heat only (Torfs 2007).

---

[54] To give a tangible example, the authors in (Asada 2003) suggest the *flanks of the finger* as the application sites of the *light source and sink* rather than the *dorsal and palmar sides of the finger*. The former locations are desirable, because both flanks have a thin epidermal layer (i.e., less attenuation of the penetrating light) and, on the other hand, the lateral (digital) arteries in the finger would necessarily reside in the resulting *light pathway*; see Fig. 5.9a for the location of the digital arteries.

[55] In analogy with Footnote 53—comparing the finger and earlobe—there are distinct differences in *optic biosignals* recorded from the *finger* or the *chest* (Kaniusas 2006). In contrast to the fingertip, the chest contains less defined *composition of tissue layers* under the skin surface, includes usually a thick layer of the *subcutaneous adipose tissue*, and, more importantly, is relatively *weakly perfused* by blood. Thus large variability in the resulting light *probing depth* as well as the (estimated) absorption and scattering properties can be expected from one subject to another; compare Fig. 5.24.

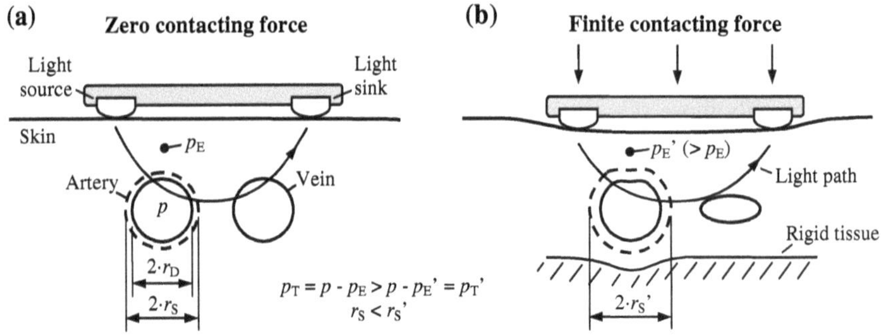

Fig. 5.26 Phenomenological effects of the contacting force (between the optical sensor in the reflectance mode and the application site of the sensor) on arterial and venous vessels below the skin surface; compare Fig. 5.27. (**a**) Absent contacting force with indicated transmural pressure p_T (= $p - p_E$) across the arterial wall, the external pressure p_E outside the artery, and the blood pressure p within the artery (Fig. 2.44b). Here r_D and r_S are the end-diastolic and end-systolic artery radius, respectively (Fig. 5.14c). (**b**) Finite (relatively weak) contacting force increases the volume pulsation of the artery on the assumption that $p_E < p_E' \leq p$, i.e., the pressures $p_T > p_T'$ satisfy $p_T \geq 0$ and $p_T' \geq 0$. The real proportions of the sensor and vessels are neglected

Contacting Force and Skin Temperature

Contacting Force

It is important to note that the *contacting force* between the optical sensor and its application site (on the skin) *affects* significantly the resulting *optic biosignal* (Teng 2004); compare section "Motion Artefacts" in Sect. 5.1.2.3 and Footnote 53. As illustrated in Fig. 5.26, the pressure exerted on the skin by the sensor (in the reflectance mode, Fig. 5.22b) can deform the geometry of (superficial) arterial and venous vessels, flatten the vessel wall, and can (significantly) reduce the diameter of affected vessels.

As the result of *increasing contacting force*, the pressure outside of the *arterial vessels* (located nearby and beneath the application site) increases and thus approaches the intra-arterial pressure of vessels. Consequently, the *transmural pressure* p_T—the difference between the intra-arterial pressure and the pressure outside of the blood vessel (Footnote 141 in Sect. 2)—decreases from positive values towards zero ($p_T \geq 0$) and the mechanical *stress within the arterial wall* is progressively relieved. Because of the relieved stress, the *arterial compliance* increases (compare Fig. 2.42 for low pressure values) and thus the corresponding *volume pulsations* increase.

Figure 5.26 illustrates the decrease of p_T from zero contacting force (Fig. 5.26a) to finite contacting force (Fig. 5.26b), whereas the *volume pulsation* over the cardiac cycle increases; compare *systolic radii* r_S in Fig. 5.26a, b. In other words, a slight pressure on the skin—up to a certain point with $p_T \geq 0$—tends to increase the volume pulsation and thus the *pulsatile (alternating) component* of the optic biosignal, the *optoplethysmogram* (i.e., the pulsatile part of the light absorbance, $\propto 1/I_{AC}$,

Fig. 5.15c). The *proximal arterial wall may flatten*, as shown in Fig. 5.26b; compare also with the tonometric method from Sect. 3.1.3.1. In fact, this pulsatile component reaches a maximum when p_T approaches zero because the *arterial compliance* peaks at $p_T = 0$. The *contacting force* yielding the maximum amplitude of the pulsatile component is thus near the *mean arterial pressure*[56] (of about 100 mmHg, see section "Pulse Waveforms of Pressure and Flow" in Sect. 2.5.2.3); compare with *external pressures* which different vessels can still sustain, see section "Motion Artefacts" in Sect. 5.1.2.3. From a practical point of view, such *local pressurization* can be applied to *amplify the pulsatile component* of the optoplethysmogram (Asada 2003).

After the peak point in the pulsatile deflection amplitude (at $p_T = 0$), the *volume pulsation* in arterial vessels and the corresponding *pulsatile component* of the optoplethysmogram begin to decrease with still increasing *contacting force* because of the progressing *arterial occlusion*. Eventually, an affected artery is pushed against rigid distal tissues (or bone) behind this artery (Fig. 5.26b), so that the *distal arterial wall may completely flatten*, causing the *volume pulsation of this artery to disappear*.[57]

It should be recalled that *venous vessels* are usually larger and more compliant than arterial vessels, so that veins are *more susceptible to collapse* by the contacting force; compare section "Motion Artefacts" in Sect. 5.1.2.3. With *increasing contacting force* (e.g., on the skin of the fingertip), the blood flow in veins in the region of compressed tissue becomes *constricted* and the diameter of veins decreases, as demonstrated in Fig. 5.26b. When this force reaches a certain level, the affected veins become completely *squeezed*. Further increase in the force renders the compressed *tissue bloodless* (e.g., the force pushes all blood out of the fingertip vessels). In addition, *venous return of blood* (e.g., in the fingertip) is progressively restricted with increasing contacting force, so that blood is *pooled in the capillaries* upstream from venules (e.g., in the capillaries under the fingernail (Mascaro 2001)). In optical terms, the effective *volume of blood* in the transilluminated region under the *skin* decreases with increasing contacting force. Consequently, the light absorption is reduced and the *non-pulsatile (direct) component* of the optoplethysmogram (i.e., the non-pulsatile part of the light absorbance, $\propto 1/I_{DC}$, Fig. 5.15c) decreases accordingly. There are indications that the non-pulsatile component approaches an *asymptotic value* when the affected veins are completely collapsed (Teng 2004). In addition, the aforementioned *arterial occlusion* contributes to

---

[56] It should be noted that such a *high contacting pressure*—namely, close to the *mean arterial pressure*—can not be applied on the skin for a long period of time in order to amplify the pulsatile component (Asada 2003). It would completely squeeze capillary beds and venous vessels (section "Motion Artefacts" in Sect. 5.1.2.3), thus limiting and *impeding the supply of arterial blood and the return of venous blood.*

[57] Similar behaviour of the pulsatile amplitude in the arterial vessel was already observed in the *oscillometric method* for the monitoring of *blood pressure* (Sect. 3.1.3.1). Here *radial oscillations of the arterial vessel wall* reach their maximum amplitude when the cuff pressure (i.e., the external pressure outside the vessel, Fig. 5.26) passes the *mean arterial pressure.* As the cuff pressure increases above or decreases below the mean blood pressure, volume pulsations of the arterial vessel decrease.

reduced volume of blood in the transilluminated region and thus to the diminished non-pulsatile component.

From an experimental point of view, the *pulsatile component* of the optoplethysmogram was reported to increase first and then decrease with increasing *contacting force* (from 0.2 to 1.8 N on the skin of the fingertip in accordance with Teng (2004)). The maximum amplitude of the pulsatile component was found within the range of 0.2–0.4 N. Own data (Fig. 5.27d) and results in Mascaro (2001) confirm that the *non-pulsatile component* of the optoplethysmogram in the *reflectance mode* decreases with increasing force. Interestingly, the *ratio of the pulsatile to non-pulsatile component* was reported to be affected less by changing contacting force than the pulsatile component alone or the non-pulsatile component alone (Teng 2004). It indicates a favourable *robustness* of this ratio for diagnostic purposes, as well as of the discussed *ratio* $\mathcal{R}$ for the experimental assessment of blood oxygenation (5.17).

The *normalized pulse area*, i.e., the area of a single cardiac pulse in the optoplethysmogram normalized by the period and deflection amplitude of this particular pulse, was reported to decrease with increasing *contacting force* (Teng 2004). Because of this particular normalization, the resulting area changes can be ascribed to *waveform changes of the optoplethysmogram* within a single cardiac cycle (as caused by varying contacting force). Namely, individual *cardiac pulses become sharper with increasing contacting force*, which is in line with the diminished normalized pulse area.

The discussed changes of the *pulsatile and non-pulsatile components* are illustrated in Fig. 5.27. With increasing *contacting force* (Fig. 5.27a), the *pulsatile component* of the optoplethysmogram increases temporarily, i.e., from the *initial pulsation* (the deflection $s_{\mathrm{S,D}}$ at the time A in Fig. 5.27c), to the *maximum pulsation* ($s_{\mathrm{S,D}}$ at the time B), and to the *decaying pulsation* ($s_{\mathrm{S,D}}$ at the time C). It can be observed that the pulsation disappears almost entirely at higher force levels. As expected, the *non-pulsatile component* of the optoplethysmogram initially decreases with (slightly) increasing contacting force and then seems to level off (Fig. 5.27d). Interestingly, after the contacting force has been removed, the non-pulsatile component exhibits a strong peak (at the time D in Fig. 5.27c). This peak could be due to strengthened *flush back of venous blood* after this blood has been pressed out of the affected tissue region (and pooled upstream in the capillaries). A comparison of pulses at the times A and C in Fig. 5.27c confirms that *cardiac pulses become sharper with increasing contacting force.*

There are indications that the *pulse wave velocity* (2.22) decreases along the arteries affected by the *contacting force* (Teng 2004). This can be explained by the fact that the stress within the arterial wall is progressively relieved with increasing contacting force and—as already mentioned—the effective stiffness of the wall decreases (see section “Pulse Propagation” in Sect. 2.5.2.3). An indirect confirmation of the latter behaviour could be found in Fig. 5.27c.

A pulse at the time C (Fig. 5.27c) yields a clear secondary wave due to *pulse wave reflection*, i.e., it yields a *reflected wave* after the primary peak (or after the incident *forward wave*); see section “Reflected Pulse Propagation” in Sect. 2.5.2.3.

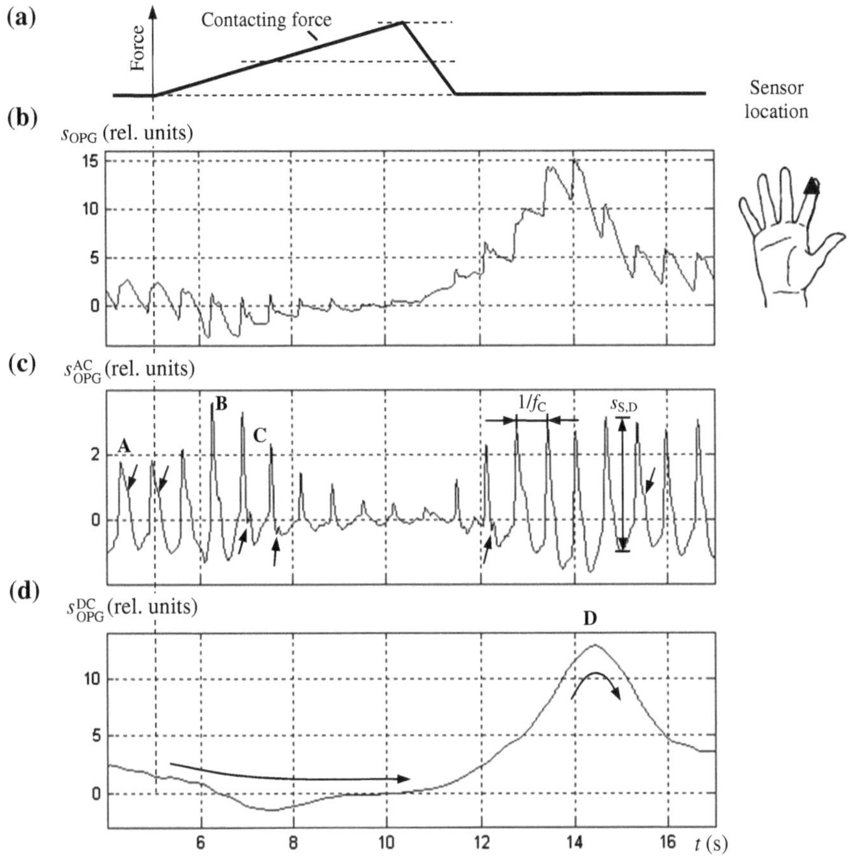

Fig. 5.27 Experimental effects of the varying contacting force between the optical sensor (in the reflectance mode, Fig. 5.22b) and its application site on optic biosignal optoplethysmogram s_{OPG} (from a fingertip on the right hand); compare Fig. 5.26. (**a**) Qualitative temporary increase of the contacting force. (**b**) The total biosignal s_{OPG} (= s_{OPG}^{AC} + s_{OPG}^{DC}). (**c**) The pulsatile (alternating) component s_{OPG}^{AC} of s_{OPG}, which oscillates with indicated heart rate f_C. The component s_{OPG}^{AC} was isolated from s_{OPG}, by a high-pass filter with the cut-off frequency of 0.4 Hz; compare Footnote 38. The systolic-diastolic deflection $s_{S,D}$ of s_{OPG} is indicated (Fig. 5.15a). (**d**) The non-pulsatile (direct) component s_{OPG}^{DC} of s_{OPG} isolated by a low-pass filter with the cut-off frequency of 0.2 Hz. Approximate starts of the reflected waves (e.g., inflection points around the time A) and the reflected waves themselves (e.g., around the time C) are indicated by *arrows*; compare Figs. 3.31c, d and 3.36

In contrast, a pulse at the time A yields only an *inflection point* after the primary peak, which usually denotes an approximate start of the reflected wave (Fig. 2.48). That is, the diminished stress in the arterial wall reduces the pulse wave velocity, delays the arrival of the reflected wave at the transilluminated site, and thus renders this *late reflected wave* visible in the diastolic phase (at the time C but not A in Fig. 5.27c); compare Footnote 40. In contrast, the initial pulsation (before force application) yields an *early reflected wave* which is almost merged with the forward wave (at the time A in Fig. 5.27c), whereas the presence of this reflected wave manifests only as the inflection point after the systolic peak.

As an additional *minor impact* of varying *contacting force*, the transilluminated region of tissue under the skin can be expected to reach deeper *tissue layers* with increasing contacting force. This is because the skin is slightly deformed and the (superficial) tissues are slightly compressed (Fig. 5.26). Since deeper tissue layers are better perfused, it results in a slight increase of the local light absorption (compare Fig. 5.24) and thus in a corresponding change of the *pulsatile and non-pulsatile components* of the optoplethysmogram.

Skin Temperature

The local temperature of the skin (and the superficial tissue) where the optical sensor is applied also impacts the optic biosignal. For instance, decreasing skin temperature is usually related to peripheral *vasoconstriction* of vessels, e.g., in fingers subjected to a cold test with the optical sensor on the fingertip. Consequently, the *volume pulsation* of transilluminated vessels decreases so that the *pulsatile component* of the optoplethysmogram diminishes significantly (Teng 2004). The *non-pulsatile component* of the optoplethysmogram can also be expected to decrease due to *reduced blood volume* and thus due to *reduced non-pulsatile absorption* of light in tissue.

Figure 5.28 illustrates changes of the *pulsatile and non-pulsatile components* in the course of a *local cold test*. In this test, the left hand was temporarily immersed into (relatively) cold water while the optic biosignal was continuously recorded from the finger on the other hand, the right hand. It can be expected that vasoconstrictive effects in the left and right hands are mutually interrelated because *sympathetic vasoconstrictive stimuli* originate in the same central nervous system. It can be observed in Fig. 5.28 that—with the hand in the cold water (starting at about 31 s in Fig. 5.28a)—the *pulsatile component* of the optoplethysmogram begins to decrease, i.e., the deflection $s_{\mathrm{S,D}}$ decreases in Fig. 5.28c. When the hand is taken out of the water (at the time of about 36 s), the deflection amplitude of the pulsatile component recovers, i.e., $s_{\mathrm{S,D}}$ increases again in Fig. 5.28c. As expected, the *non-pulsatile component* of the optoplethysmogram decreases temporarily (Fig. 5.28d). In addition, multiple cold tests on different persons have revealed another interesting observation. When the hand was kept in the cold water for an extended time period of at least 20 s, the recovery of both the pulsatile deflection amplitude and the level of the non-pulsatile component (i.e., their increase towards respective initial levels) had started and even completed before the hand was taken out of the cold water. This behaviour could indicate active regulatory mechanisms in the periphery.

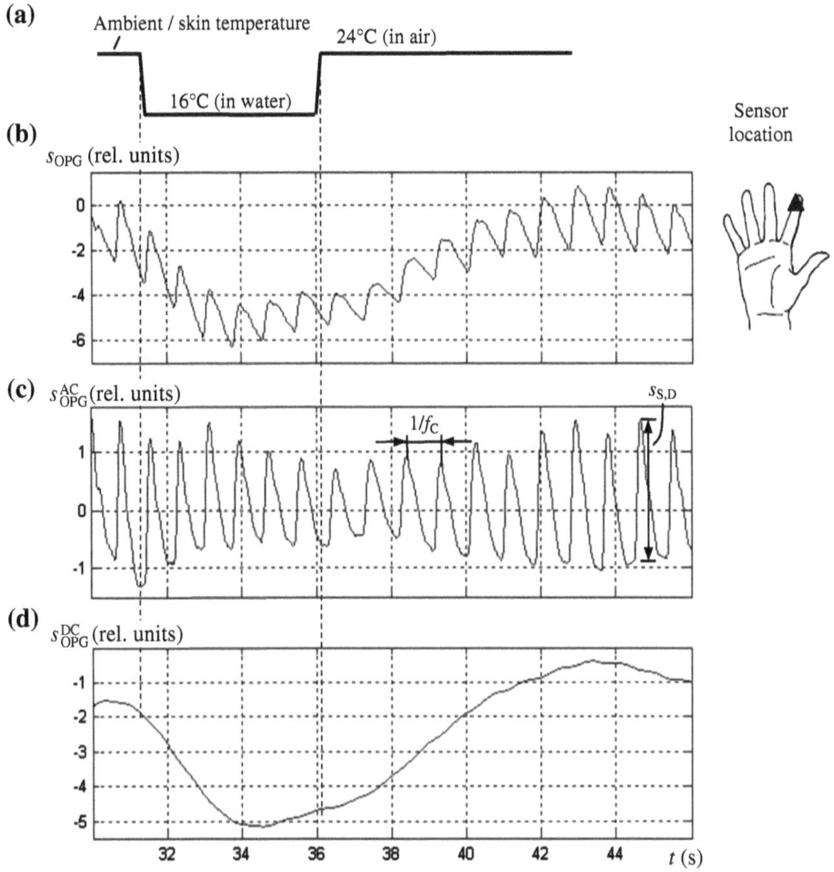

Fig. 5.28 Experimental effects of a local cold test on optic biosignal optoplethysmogram s_{OPG} (from a fingertip on the right hand) with the optical sensor in the reflectance mode (Fig. 5.22b). (**a**) Qualitative temporary decrease of the ambient temperature of the left hand which was temporarily immersed into cold water. (**b**) The total biosignal s_{OPG} $(= s_{OPG}^{AC} + s_{OPG}^{DC})$. (**c**) The pulsatile (alternating) component s_{OPG}^{AC} of s_{OPG}, which oscillates with indicated heart rate f_C. The component s_{OPG}^{AC} was isolated from s_{OPG} by a high-pass filter with the cut-off frequency of 0.4 Hz; compare Footnote 38. The systolic-diastolic deflection $s_{S,D}$ of s_{OPG} is indicated (Fig. 5.15a). (**d**) The non-pulsatile (direct) component s_{OPG}^{DC} of s_{OPG}, isolated by a low-pass filter with the cut-off frequency of 0.2 Hz

5.2.1.3 Light Sink

The *light sink* contains typically a *photodiode* used to register the *transmitted light intensity*. Thus an *induced optic biosignal* is established (Fig. 5.1a), which is inversely related to the transmitted light intensity or, in other words, is proportional to the *light absorption strength* (Fig. 5.15a).

As illustrated in Fig. 5.5b, the photodiode is made of semiconductor materials that *detect light* at the *pn junction* (Footnote 7). In contrast to the light-emitting diode as a light source (Fig. 5.5a), the photodiode is *reverse-biased* by an *external voltage*, i.e., the positive electrode is connected to the n-side (*cathode*) and the negative electrode is connected to the p-side (*anode*). The effective width of the *charged layer*—with a build-in electric field (Sect. 5.1.1.2)—increases with the applied external voltage. Likewise, the (isolating) *depletion layer* without free charge carriers widens in comparison to an unbiased pn junction (without any external voltage). Ideally, there is no current across the pn junction because negative electrons are attracted by the positive electrode and positive holes are attracted by the negative electrode (Fig. 5.5b). In contrast to the light-emitting diode (Fig. 5.5a), both electrons and holes do not meet at the pn junction (Fig. 5.5b).

In the presence of a *light photon* with an appropriate energy (5.3) in the pn junction, namely, in the (widened) *depletion region*, an *electron-hole pair* is generated in this region (known as *inner photoelectric effect*). That is, the *absorbed photon* brings energy to the atom it collides with. This energy *elevates an electron* of the affected atom from its *valence band* into its *conduction band* (Footnote 6). The electron is freed and can become part of a (current) conduction process, see below. The original slot of this elevated electron left behind in the valence band is an *electron hole*, which behaves as a positively charged particle (Footnote 7). The electron-hole pair is generated provided that the photon energy *exceeds the energy gap* W_g (5.3) between the (wide) *valence band* and (wide) *conduction band* (Footnote 5); compare Sect. 5.1.1.2.

The *free electron* is created in the conduction band which then is swept from the depletion layer by the built-in *electric field in the pn junction* towards the positive electrode (Fig. 5.5b). That is, this free electron is swept against the concentration gradient of charges (compare Sect. 2.1.2.2). At the same time, the *created hole* in the valence band is attracted towards the negative electrode, so that an electric *current starts to flow* (*photocurrent*) across the junction in response to the light photon (known as operation with *reverse bias*, Fig. 5.5b).

It should be noted that some generated electrons and holes may *recombine* directly in the *pn junction*, i.e., annihilate each other because of finite *carrier lifetime* (compare Sect. 5.1.1.2). Such direct recombination would decrease the number of created free electrons and holes in the pn junction; however, this recombination process can be usually neglected because of the relatively long carrier lifetime (> 1 μs) related to the time the carrier stay in the pn junction (< 100 ns).

The *lower bound of the photon energy* detected by the photodiode is given by W_g, i.e., the *largest wavelength* λ_g of light which can be detected amounts to $v \cdot h/W_g$ (5.3); compare Sect. 5.1.1.2. Furthermore, the *upper bound of the photon energy* (> W_g) detected by the photodiode, i.e., the *smallest wavelength* (< λ_g), is mainly determined by the corresponding *penetration depth* of light into the pn junction (or into silicon). The penetration depth of a photon into silicon is wavelength dependent; namely, this depth decreases with decreasing λ (< λ_g). Photons with an energy below W_g (or with $\lambda > \lambda_g$) transverse the pn junction (or silicon) without any interaction, the semiconductor material is transparent for this light (compare Footnote 13).

That is, the induced *photocurrent* in the *photodiode* is a proportional measure of the *transmitted light intensity* through biological tissue (as detected by the light sink).

The transmitted light intensity—as an approximation for small changes in the light absorption in (5.4) (small signal approximation)—is inversely related to the *induced optic biosignal* (proportional to the *light absorption*). As illustrated in Fig. 5.1a, the optic biosignal *optoplethysmogram* constitutes the output signal of the *optical sensor*.

5.2.1.4 Adverse Health Effects and Exposure Limits

Since optic biosignals comprise *induced biosignals*, i.e., an artificial light is induced into biological tissue (Fig. 5.1), *adverse health effects* (section "Health Effects" in Sect. 5.2.1.4) and applicable *limits of this artificial exposure* (section "Remarks on Exposure Limits and Optic Biosignals" in Sect. 5.2.1.4) should be considered. In particular, the exposure of the skin is relevant, where the optical sensor is applied on. Public exposure limits of *visible and infrared incoherent radiation*—as used in optical sensors (Sects. 5.1.1.2 and "Light Wavelength" in Sect. 5.2.1.2)—incident on the eye or skin are provided in ICNIRP (1997), where irradiance levels (incident optical power per illuminated area) and exposure durations are known or controlled. It should be stressed that

- the *eye* and
- the *skin*

are the organs which are most *susceptible to adverse effects* induced by optical radiation (section "Organs at Risk—Eye and Skin" in Sect. 5.2.1.4). The eye is normally more sensitive to injury from visible and infrared radiation than the skin. Furthermore, the consequences of an overexposure of the eye are generally more serious.

Health Effects

A *strong wavelength dependence* applies to possible damage mechanisms and the corresponding thresholds applicable to the *eye*, i.e., its different parts such as cornea, lens, and retina, and the *skin*. Therefore, exposure to *broadband and incoherent light sources* (Sect. 5.1.1.2) emitting light in a wide band of the frequency spectrum must be evaluated against possible *adverse effects* due to

- *photochemical interactions* and
- *thermal interactions*.

Photochemical interactions—as introduced in Footnote 15—are subjected to the *principle of reciprocity*.[58] That is, an *inverse relationship* exists between *light*

[58] In contrast to *thermal interactions*, *photochemical interactions* are subjected to the *principle of reciprocity*. It implies that the threshold for photochemical injury is proportional to the *total exposure*, the product of

- *light intensity* (such as brightness of the light source or irradiance of a surface) and
- *exposure duration*.

intensity and *exposure duration* that determines the *threshold* for photochemical injury. It should be stressed that photochemical interactions depend strongly on the wavelength (see action spectra in section "Remarks on Exposure Limits and Optic Biosignals" in Sect. 5.2.1.4). In particular,

- *visible radiation* is subjected to an important *photochemical interaction*, known as *blue-light retinal injury* or *photoretinitis*. It can occur in the range of 300–700 nm, i.e., in the range of shorter wavelengths of visible light, whereas this interaction peaks at about 440 nm for the intact eye.
- In contrast, photochemical effects are not likely to be induced by *infrared radiation* because the corresponding photon energy (5.3) is not sufficient to induce chemical reactions or to ionize structures (Footnote 15).

There is a potential *enhancement of photochemical hazards* with increasing *tissue temperature.*

Thermal interactions mean generation of *heat* in tissues in response to the absorption of light; for details see section "Volume Effects" in Sect. 5.1.2.2 and Sect. 6. In contrast to photochemical interactions, thermal interactions are not subjected to the principle of reciprocity (Footnote 58). In other words, there is *no inverse relationship* between the incident *light intensity* and the corresponding *exposure duration* that determines the *threshold* for thermal injury. This is because thermal injury is strongly dependent upon *heat conduction* away from the irradiated spot of tissue by *active regulatory mechanisms*; see Footnote 17 and Sect. 3. Obviously, *passive mechanisms* of heat conduction are also involved such as heat diffusion from regions of higher temperature to regions of lower temperature. That is, an *intense exposure within a short period of time*, typically within seconds, is needed to cause *thermal damage* (e.g., tissue coagulation). When the exposure is less intense, heat is conducted away from the exposed spot via surrounding tissue and thermal damage is thus prevented.

In addition, the resulting cooling (or *radial heat flow*) is more efficient for *small irradiated spots* (e.g., in the retina of the *eye*) so that thermal injury of relatively small (retinal) spots requires higher irradiances than thermal injury of irradiated larger (retinal) spots. For instance, the *threshold for the thermal retinal injury* varies more or less inversely with the diameter of the retinal spot; this relationship is only applicable for the retinal image diameters from 20 μm to 2 mm (ICNIRP 1997).

In analogy, large areas of the *skin* will be more readily heated to higher temperatures than smaller areas provided the *irradiance level* incident on the skin is the same (ICNIRP 2006). For superficial *thermal burn* of biological tissue, an absolute temperature of *at least 45 °C* is necessary (e.g., in the *retina or skin*); even higher

(Footnote 58 continued)

For instance, *blue-light retinal injury* can result from viewing either a bright light for a short time or a less bright light for a longer time.

Obviously, the same principle of reciprocity applies in *photography*. The effect of light on the film is proportional to the product of light intensity (determined by aperture of an optical system) and exposure duration (determined by shutter speed).

temperatures are required for thermal burn provided the exposure is relatively short (ICNIRP 1997). For comparison, a *tolerated increase* in the core body temperature is limited to only *about 1 °C* (i.e., limited to the absolute core body temperature of 38 °C) to *avoid heat stroke*; see Sects. 3.1.5 and Sect. 6. Obviously, increasing *ambient temperature* and increasing *initial temperature* of the exposed skin raise risks of thermal injury of the *eyes and skin.*

Thermal effects can be induced by both *visible and infrared radiation* (section "Volume Effects" in Sect. 5.1.2.2) and can be subdivided into:

- thermal injury of the *retina*, affective in the spectral range of about 380–1,400 nm;
- thermal injury of the crystalline *lens* (e.g., cataract), affective in the range of about 780–3,000 nm;
- thermal injury of the *cornea* (e.g., clouding of the cornea), affective in the range of about 1,400 nm–1 mm; and
- thermal injury of the *skin* (e.g., thermal burn and erythema ab igne), affective in the approximate range of 380 nm–1 mm.

Generally, light photons with shorter wavelength (and higher photon energy, (5.3)) are more biologically active. For *short exposure* durations less than a few seconds in the *retina*, adverse *thermal effects* usually dominate (ICNIRP 1997), provided radiance limits of the retina are exceeded (section "Remarks on Exposure Limits and Optic Biosignals" in Sect. 5.2.1.4). Adverse *photochemical effects*, rather than thermal effects, dominate in the retina in the *visible region below 600–700 nm* for *longer exposure* times in excess of 10 s. In the *infrared region, thermal effects* in the retina still prevail for exposure times longer than 10 s; photochemical effects are absent for infrared light. Concerning the *skin*, thermal effects dominate in both visible and infrared regions. Obviously, photochemical and thermal effects may occur at the same time, e.g., photoretinitis and thermal injury of the lens, given radiant exposure above limits.

Organs at Risk—Eye and Skin

The human eye and skin deserve particular attention as *target organs* of visible and infrared radiation.

Eye

In the *eye*, to begin with, the *cornea, lens, and retina* can be at risk in terms of *photochemical and thermal effects* when the eye is subjected to *visible* and *near-infrared light* (section "Health Effects" in Sect. 5.2.1.4). The special thing about the exposure of the eye is that its *natural aversion response* (natural *reflex response*) to bright light reduces substantially the potentially hazardous exposure. This optical aversion response *limits the duration of exposure* to less than 0.25 s from viewing bright light, protecting the eye against injury (photochemical and thermal injury).

Mechanisms involved in the *optical aversion response* comprise (continuous) *involuntary eye movements*, eye-blink reflex, the degree of lid closure, and obviously *voluntary eye movements*. In addition, *eye movements spread the incident optical energy* (of relatively small images) on the retinal surface for exposure durations > 0.25 s. Consequently, retinal areas affected by the viewed image are larger than the actual image size on the retina. This *time-averaging* diminishes effective *irradiance of the retina* (i.e., incident optical power per retinal area) for exposure durations > 0.25 s providing *natural protection of the eye*.

It is important to stress that only *visible light* yields the *optical aversion response* in the eye. This protective response is missing when the human eye is exposed to *infrared light* only.[59] The reason for this is that *photochemical effects* are not induced by infrared light, in contrast to visible light (section "Health Effects" in Sect. 5.2.1.4). However, *heat* impact on the cornea from ambient infrared light may also induce a relatively slow (thermal) aversion response.

Skin

The human *skin* can be at risk in terms of mainly *thermal effects* when subjected to visible and near-infrared light (section "Health Effects" in Sect. 5.2.1.4). However, the *thermal discomfort*, i.e., whole-body heat stress or even pain response, sensed by the skin (or the cornea) will also generate a *natural aversion response* if the total *effective irradiance*[60] is sufficiently high. This *thermal aversion response limits the duration of thermal exposure* to a few seconds, i.e., to exposure duration < 10 s according to ICNIRP (2006). Interestingly, the sensation of *heat* by thermal (pain) sensorial endings (Fig. 2.21a)—and thus the corresponding thermal aversion response—is more certain when the *irradiated area of the skin* is larger (ICNIRP 2006). Because of the thermal aversion response and the aforementioned *regulatory mechanisms* of heat conduction (section "Health Effects" in Sect. 5.2.1.4), only very high and very *brief irradiances* pose a thermal hazard to the skin; unless irradiation extends over a relatively long period of time (> 10 s, see Fig. 5.29).

The *penetration depth* of light incident on the skin determines the *spatial distribution of the induced heat* under the skin surface. This depth is reciprocal of the effective *attenuation coefficient* α which accounts for both *absorption and scattering properties* of tissue; see Sects. 5.2.1.1 and (5.10). Since these properties are

[59] The *exposure to infrared light blended with visible light* is less risky than the exposure to infrared light only. This is because the *optical aversion response* is due to visible light only.

[60] It should be noted that the *effective irradiance* is the difference between the irradiance *on the skin* coming from external sources and the irradiance emitted *from the skin* itself (according to *thermal black-body radiation*, see Footnote 3). As noted in Sect. 3.1.5, the *radiation emission from the human skin* plays a major role at *room temperature* when there are no external optical sources of high temperature (≫ 30 °C, i.e., much higher than the approximate skin temperature, Fig. 3.21a). If an external optical source is present and its radiation source's temperature (e.g., of several hundred degrees Celsius) is much higher than 30 °C, the *radiation emission from the skin can be neglected*. Then the effective irradiance equals the irradiance from the external source only.

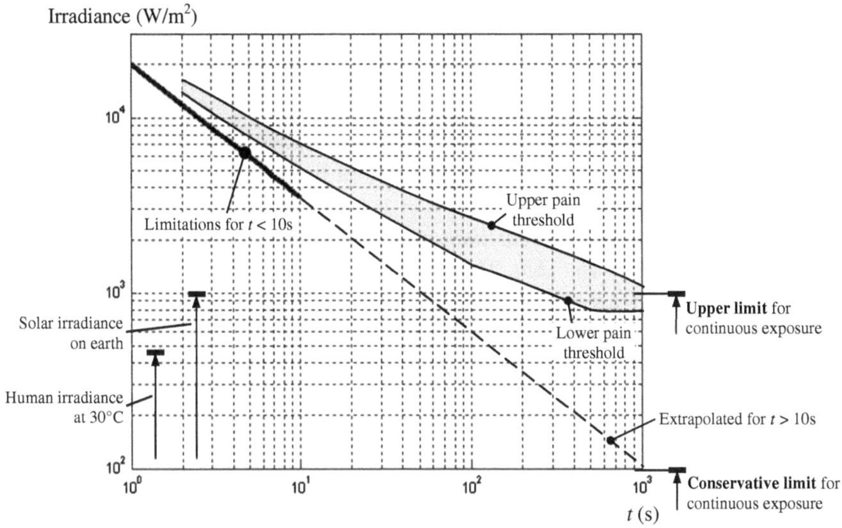

Fig. 5.29 Limitations on irradiance exposure of the human skin to visible and infrared radiation (*thick solid line*) to prevent thermal injury to the skin. Limitations are given for exposure durations *t* for up to 10 s in accordance with the guidelines (ICNIRP 1997, 2006). For comparison, pain thresholds are depicted as a function of *t*, where the *grey area* indicates the spread of thresholds due to individual variations, as adopted from (ICNIRP 2006)

wavelength dependent (Sect. 5.1.2.2 and Fig. 5.7), the penetration depth also strongly *depends on the wavelength* of the incident light. Figure 5.23 demonstrates *reported penetration depths* over the wavelength of *visible and infrared radiation* incident on the *human skin*. As discussed in section "Penetration Depth" in Sect. 5.2.1.1, the depth increases in the visible range with increasing wavelength and then strongly decreases in the infrared range. The relatively long infrared wavelengths are only absorbed by the outermost layer of the skin, the *epidermis* (Fig. 5.23). That is, only the *epidermis* is directly heated by radiation with *long infrared wavelengths*. In this case, *heating of deeper layers* is only achieved by heat transfer inwards, i.e., by *passive heat conduction* and *blood flow* in tissues; see section "Health Effects" in Sect. 5.2.1.4 and Footnote 17.

The optical *reflectance* of the skin also plays an important role with respect to the impact of external radiant sources; compare section "Inhomogeneity Effects" in Sect. 5.1.2.2. As illustrated in Fig. 5.23, the *skin reflects* predominantly

- *visible radiation* (including solar radiation) and
- *infrared radiation* within the range of 780–1,400 nm (ICNIRP 2006).

By contrast, *infrared radiation* with the wavelengths > 1,400 nm (e.g., dominant in hot industrial work environments) is only weakly reflected by the skin and heats superficial layers of the skin.

In fact, thermal injury of the *skin* is highly dependent on *exposure duration*, the *size of the irradiated spot* (or of the *source size*), the *ambient temperature*, and the *initial skin temperature*. For instance, the distal *skin* temperature is usually lower than 30 °C (according to Fig. 3.21a); if compared with 37 °C of the *retina*, it is clear that the retina is exposed to higher risk of thermal damage than the skin. In analogy, a high ambient temperature favours heat strain of the body at a given level of skin irradiance.

Figure 5.29 illustrates *current limitations* of *visible and infrared radiation* to avoid *thermal injury to the skin*. That is, the *allowed irradiance* (in W/m^2) of the *unprotected skin* decreases with increasing *exposure duration* (for up to 10 s) because the efficiency of the discussed *regulatory mechanisms* holds only for a limited time. In other words, an additional *heat stress*—which can still be maintained (sustained) by the skin and body—for another period of time of uninterrupted exposure becomes less and less with increasing *exposure duration*. Consequently, the current guidelines in (ICNIRP 1997, 2006) limit the total radiant exposure (in J/m^2, i.e., incident optical *area-related energy* but not *power*) by the non-linear factor $t^{1/4}$, where *t is the exposure duration*. This factor implies a *saturation effect* (for the incident area-related energy), whereas a linear factor such as t^1 does not saturate. *No general limitations* of irradiance of the skin are provided by the above guidelines for *exposure durations greater than 10 s* because the *threshold irradiance* for thermal injury (thermal burn) depends strongly on ambient conditions, the initial skin temperature, and thermal aversion response (see above).

For comparison, Fig. 5.29 depicts the region of the *effective irradiance* (Footnote 60) above which *thermal pain onsets*. The latter region applies to the *unprotected skin* irradiated by thermal radiation for *short and extended periods of time*, i.e., for both shorter and longer than 10 s. According to ICNIRP (2006), these shown pain thresholds are independent on the size of the irradiated skin area and cover both *visible and infrared radiation*. It can be observed that the irradiance strength that just produces a pain sensation decreases with increasing *exposure duration*. The pain thresholds are depicted as a (grey) region, a spread of values due to individual variations of the threshold level. As expected, the general *exposure limits* (for up to 10 s) are well *below the pain thresholds*. In addition, the extrapolated limits for exposure durations longer than 10 s reside also below the pain thresholds (Fig. 5.29).

The above *pain thresholds* (Fig. 5.29) provide a *practical guide* for the limitation of irradiance to avoid thermal injury (burn) to the skin with *exposure durations greater than 10 s*. This is because *thermal pain* is usually induced by skin temperatures that are still lower than the temperatures needed to produce a *thermal burn*. In other words, thermal pain usually precedes thermal burn. Consequently, the provided pain thresholds in Fig. 5.29 limit external optical irradiance to safe levels that prevent thermal burn.

Generally speaking, a very *conservative irradiance limit* for *continuous exposure of the whole body* to *visible and infrared radiation* is *100 W/m^2*—overly conservative for cooler environments but rather applicable for very warm environments—as can be derived from the current guidelines in ICNIRP (1997, 2006). Conversely, an *upper irradiance limit for continuous exposure* of the whole body—

even in cooler environments—is *1 kW/m²*; compare with the absolute levels of irradiance from Fig. 5.29.

Additional insights into the tolerated levels of irradiance by the skin offer their comparison with the typical incident *solar irradiance on the earth surface* (Fig. 5.29). Interestingly, the maximum solar irradiance is about 1 kW/m^2 which already equals to the (thermal) pain thresholds for exposure durations in excess of about 5 min. On the other hand, the power emitted by the human body as infrared radiation (*human irradiance*) is about 480 W/m^2 and as such is denoted in Fig. 5.29 for comparison aims. Here the human body is approximated as the ideal *black-body radiator* (Footnote 3) with an assumed *skin* temperature of 30 °C.

Provided *chronic exposure of the skin* is present, e.g., within the scope of *long-term registration of optic biosignals*, applicable limits for the tolerated irradiance decrease with increasing *thermal loading* of the affected body. The thermal loading is given by

- *ambient conditions* (e.g., air temperature, air movement, humidity, and clothing) and
- *heat production* within the body (e.g., metabolic heat generation and heat generated by *physical activity* of the body);

compare Sect. 3.1.5. In other words, *heat stress* and the resulting *heat strain* (body's response to heat stress) depend not only on irradiance levels but also on the effective *thermal loading*. To give an example showing the impact of the thermal loading, the *irradiance level* of about 300 W/m^2 can be tolerated for 8 h *without performing any physical activity* in accordance with the norm (DIN 2011). By contrast, only about 140 W/m^2 can be tolerated for 8 h by the human body subjected to continuous *physical activity* of 200 W.

Remarks on Exposure Limits and Optic Biosignals

In order to adequately account for the *wavelength dependence of adverse health effects*, *action spectra* are defined in guidelines (ICNIRP 1997). These action spectra are used to derive either

- *biologically effective radiance* (in $W/(m^2 \cdot sr)$) to account for the *brightness* of the external light source or
- *biologically effective irradiance* (in W/m^2) to account for the *optical power* incident on a surface (e.g., skin surface).

Here biological effectiveness implies photochemical and thermal effects. *Action spectra* are applied to *spectrally weight* radiance or irradiance from an external light source along the relevant range of the light *wavelength*. In particular, action spectra rapidly change over the wavelength in the *visible region below 600–700 nm* where *photochemical effects* dominate; compare blue-light retinal injury in section "Health Effects" in Sect. 5.2.1.4. In contrast, *thermal effects* exhibit only relatively slow changes in their spectral sensitivity, i.e., slow changes in their action spectra.

Exposure limits pertaining to *photochemical interactions* can be mainly based on the total exposure (e.g., integrated radiance, see below) because of the applicable *principle of reciprocity* (Footnote 58), i.e., *exposure duration* needs not to be specified explicitly. In contrast, exposure limits of *thermal interactions* depend also on *exposure duration* and the size of the *irradiated spot* due to the aforementioned heat dissipation by *regulatory mechanisms* (section "Health Effects" in Sect. 5.2.1.4). Therefore

- *source-related* quantities such as *radiance* (in $W/(m^2 \cdot sr)$) and *integrated radiance* (in $J/(m^2 \cdot sr)$) are used for exposure limits to protect the *retina*;
- *surface-related* quantities such as *irradiance* (in W/m^2) and *integrated irradiance* (in J/m^2) are mainly used to protect the *cornea*, *lens*, and the *skin*; compare Fig. 5.29.

In addition, the apparent *visual angle*, i.e., the angle a viewed light source subtends at the eye is used to evaluate *thermal risks*. This angle is approximately given as the quotient of the light source diameter and the viewing distance. For instance, the *exposure limit* for *thermal retinal injury* is inversely proportional to the visual angle and to exposure duration (ICNIRP 1997). In other words, the smaller the *irradiated spot* in the retina and the shorter the *exposure duration*, the more efficient is the *radial heat flow* in the retina and thus the more effective is *cooling of the retina* (section "Health Effects" in Sect. 5.2.1.4).

In addition, the *minimum visual angle* (of the light source) can be defined, above which the light source can be considered as *extended source*. This minimum angle increases with exposure duration when *involuntary eye movements* dominate and thus protect the retina; compare section "Organs at Risk—Eye and Skin" in Sect. 5.2.1.4.

Considering *adverse health effects* in view of *induced optic biosignals*, as applied in *optical sensors* (section "Light Wavelength" in Sect. 5.2.1.2), it can be concluded (from sections "Health Effects" in Sect. 5.2.1.4 and "Organs at Risk—Eye and Skin" in Sect. 5.2.1.4) that

- the applied *red wavelength* at around 660 nm (Fig. 5.8) may induce only weak *photochemical interactions* in the eye. This is because 660 nm is almost out of the action spectra of blue-light retinal injury (effective in the range of 300–700 nm).
- Both applied *red and near-infrared wavelengths* at around 660 and 890 nm, respectively (Fig. 5.8), can be subjected to *thermal interactions* in the eye and skin. However,
  - the aforementioned regulatory mechanisms (section "Health Effects" in Sect. 5.2.1.4),
  - small illuminated spots of the skin,
  - small visual angle of the light sources (i.e., relatively small size of the light-emitting diode versus relatively large handling distance of the optical sensor to the operator's eyes before the sensor is applied on the skin), and
  - limited intensity of the used light sources

reduce significantly thermal risks. Nevertheless, *skin burning problems* have been pointed out during long-term applications of the optical sensor (Asada 2003); see section "General Issues" in Sect. 5.2.1.2.

5.2.2 Registration of Optic Biosignals

As soon as the transmitted light (intensity) has been coupled from tissue into the *light sink* (Sect. 5.2.1.3), the established *optic biosignal optoplethysmogram* (Fig. 5.1a) is available for signal analysis and *diagnostic purposes*. That is, the optoplethysmogram reflects the *dynamic modulation* of the propagating light (throughout biological tissues) by diverse physiological phenomena (Sect. 5.1.2.3). In particular, the *light intensity in tissue experiences modulation* by

- (relatively) fast *cardiac activity* (section "Cardiac Activity" in Sect. 5.1.2.3),
- less fast *respiratory activity* (section "Respiratory Activity" in Sect. 5.1.2.3), and
- slow changes in *blood oxygenation* (section "Blood Oxygenation" in Sect. 5.1.2.3).

In short, the *cardiac modulation* is due to (propagating) *pulsatile waves* under the optical sensor, which cause the transilluminated *blood volume* to fluctuate in synchrony with cardiac activity (Fig. 5.15a). The specific fluctuation of the transilluminated blood volume in synchrony with the respiratory activity mainly accounts for the *respiratory modulation* (Fig. 5.15b). In addition to the latter cardiorespiratory modulation, the absorption properties of *arterial blood* itself (but not primarily of the varying blood volume) are slowly *modulated by the oxygenation level of blood* (Fig. 5.8).

The following sections will demonstrate *experimental optoplethysmograms* recorded from the *finger*, as illustrated in Fig. 5.30. *Numerous physiological parameters* and events related to cardiac activity, respiratory activity, and blood oxygenation will be shown by and derived out of optoplethysmograms. Likewise, *multiparametric data* (Sect. 1.4) will be derived from a single optical sensing device. In particular, most of the shown data were recorded with the optical sensor operated in the reflectance mode; compare Figs. 5.22b and 5.30.

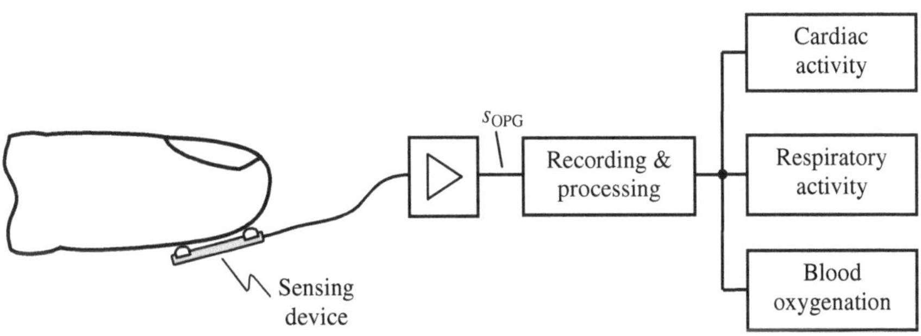

Fig. 5.30 Registration of the optic biosignal on the finger and its multiparametric processing in order to extract various physiological parameters (diagnostic parameters); compare Fig. 5.1. The optical sensor is shown in the reflectance mode (Fig. 5.22b)

Besides cardiorespiratory activity and blood oxygenation (Kaniusas 2007), *optoplethysmograms* reflect *blood pumping* and transport conditions in vessels, blood perfusion (e.g., in the superficial tissues under the skin), and *vascular compliance*; compare Fig. 3.37d (for the skin perfusion) and (5.14) (for the compliance's role in the optoplethysmogram). Moreover, optoplethysmograms have been studied for estimating total blood volume and even blood pressure (Teng 2004). For instance, Fig. 3.34 considers the impact of the stroke volume changes on the optoplethysmogram, whereas Figs. 3.36 and 5.16 consider the impact of blood pressure on the optoplethysmogram. Finally, it should be noted that optic biosignals constitute a simple and *non-invasive diagnostic approach* which has great potential in diverse clinical applications.

5.2.2.1 Cardiac Activity

Cardiac activity manifests clearly within the optic biosignal *optoplethysmogram*, as derived and discussed in section "Cardiac Activity" in Sect. 5.1.2.3. In fact, the *pulsatile alternating component* I_{AC} of the *transmitted light intensity* I reflects cardiac activity, whereas the corresponding *basic principle* is summarized in Fig. 5.14. Examples of the optoplethysmogram have already been shown in Figs. 5.15c, 5.27c,

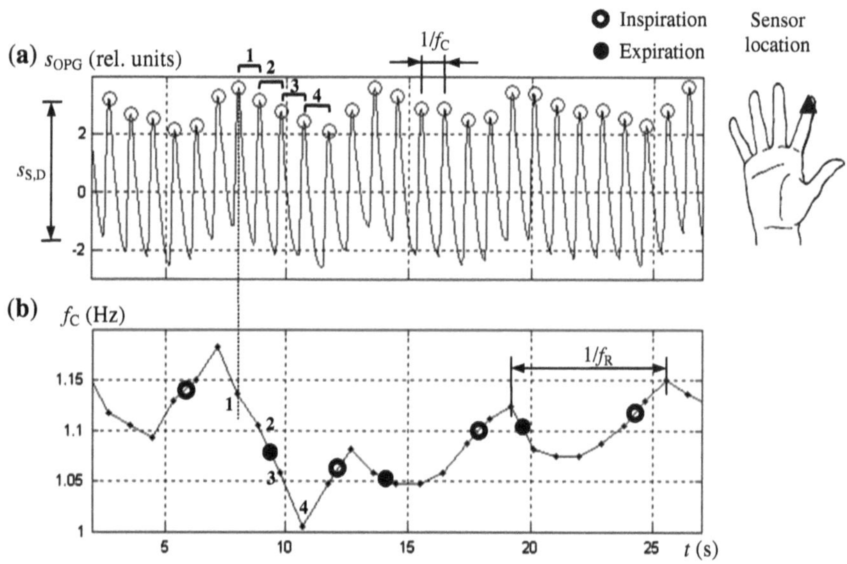

Fig. 5.31 Assessment of cardiac activity by the optic biosignal in the course of normal breathing. (**a**) Optic biosignal optoplethysmogram s_{OPG} (from a fingertip on the right hand) with the optical sensor operated in the reflectance mode (Fig. 5.22b). The systolic-diastolic deflection $s_{S,D}$ of s_{OPG} is indicated. (**b**) The instantaneous heart rate f_C derived from the time course of s_{OPG} using signal processing methods in the time domain (Footnote 61 in Sect. 4). Four pairs of s_{OPG} peaks and the corresponding values of f_C are denoted by *numbers*, whereas f_R indicates the respiratory rate

and 5.28c. Obviously, an (important) vital physiological parameter of cardiac activity is the *heart rate* f_C (Sect. 3.1.1).

Figure 5.31 demonstrates the *registration of* f_C out of the optoplethysmogram during normal breathing. In the *time domain* (Fig. 5.31a), heart beats and their periodic occurrence can be observed in the optoplethysmogram. The resulting oscillation rate is actually the rate f_C which estimated time course is shown in Fig. 5.31b. For instance, signal processing methods in the time domain can be used to derive the instantaneous f_C; see Footnote 61 in Sect. 4. It can be observed that f_C temporarily increases during inspiration, which is in line with fundamental *cardiorespiratory interrelations* (Sect. 3.2.1) and, for instance, in line with the behaviour of f_C in Fig. 3.33d. The pulsatile deflection $s_{S,D}$ is also indicated in Fig. 5.31a, as already shown in Fig. 5.15a (Footnote 37). For respiratory modulation of the optoplethysmogram, see Sect. 5.2.2.2.

As laid down in section "Cardiac Activity" in Sect. 5.1.2.3, the *pulsatile waveform* of the *optoplethysmogram* and that of *blood pressure* are quite *similar.* Both waveforms exhibit with the passing of time

- a relatively *steep systolic increase,*
- a dicrotic notch (or even multiple *maxima and minima due to reflections* of pulsatile waves, Sect. 2.5.2.3), and
- a relatively *slow diastolic decrease.*

A typical waveform of *blood pressure* is exemplified in Fig. 2.48b while that of the *optoplethysmogram* is exemplified in Figs. 5.15a and 5.32a. In optical terms, the systolic increase indicates increasing light absorption, whereas the diastolic decrease indicates decreasing absorption. It should be recalled that the *instantaneous light absorption* is directly proportional to the *instantaneous level of the optoplethysmogram.*

It should be noted that in contrast to *acoustic biosignals* (Sect. 4.1.1), *optic biosignals* exhibit smooth waveforms. For instance, *heart sounds* in Fig. 4.5a show a rather *spiky waveform* with a relatively large amount of *high frequency components* besides the basic low frequency component oscillating with f_C. For separation of these components as a function of frequency, consider Footnotes 150 in Sect. 2 and 193 in Sect. 3. The aforementioned high frequency components arise because sources of body sounds (in particular, sources of heart sounds from Fig. 4.3) emit frequency components well above the level of f_C (Sect. 4.1.1). In contrast, heart beats in the *optoplethysmogram* of Fig. 5.31a reveal a rather *smooth waveform* because *inert mechanical phenomena* are involved in the generation of the optic biosignals. As discussed in Sect. 5.1.2.3 in detail, such phenomena are related to *relatively slow* pulsatile blood shift, local changes in heterogeneous volumes (i.e., in volume fractions of blood and bloodless tissue), and to changes in blood oxygenation. The mechanisms of the sound generation, in contrast, are *relatively fast.* Likewise, the amount of *high frequency components* in the optoplethysmogram is relatively small if compared with body sounds.

Figure 5.32 illustrates an interesting case with an *ectopic beat* (or extrasystole) outside the regular sequence of heart beats, namely, a case of a *premature ventricular contraction*.[61] The optoplethysmogram (Fig. 5.32a) during the ectopic beat yields reduced deflection amplitude in comparison with the preceding and following *normal heart beats*; i.e., the *systolic-diastolic deflection* amplitude $s^2_{S,D} < s^1_{S,D}, s^3_{S,D}$. This is because an ectopic beat disrupts the (quasi) regular sequence of heart beats, so that ventricles are not completely filled by blood at the time instance of the ectopic beat (Sect. 2.4.2). Consequently, the stroke volume is reduced as well as the *pulsatile systolic-diastolic blood volume* at the site of the sensor application (5.14). That in turn reduces the deflection $s_{S,D}$ of the optoplethysmogram within the ectopic beat related to the previous and subsequent

[61] *Premature ventricular contraction* (or *extrasystole*, an additional heart beat) is a heart beat outside the regular sequence of heart beats, i.e., it is a *premature heart beat before the normal heart beat* was supposed to occur. In contrast to normal heart beat,

- this extrasystole is (*spontaneously or artificially* via *electrical stimulation*, see Sect. 6) *initiated in ventricles* (or in the Purkinje fibers, Fig. 2.35) by a pacemaker other than the sinoatrial node; compare Footnote 120 in Sect. 2. In addition,
- *slowed conduction* of the (action) impulse in ventricles may lead to a local re-excitation of ventricles, known as (single) *re-entry mechanism*. Here the impulse propagates along closed pathways (i.e., *loops* in the heart) such that the excitation wave front returns to a previously excited tissue after a certain time delay, i.e., after the *refractory period*, to be more precise, after the absolute refractory period (Sect. 2.2.2). This is long enough to permit this re-entered tissue to regain its excitability and to become re-excited. The re-entry can be promoted not only by slowing *conduction velocity* but also by shortening the *refractory period* and also by a *dilated heart*. Likewise, the wavelength of the impulse (given by the product of the conduction velocity and refractory period, compare (5.1)) must be shorter than the physical length of the aforementioned loops because the excitable tissue—into which the impulse is re-entering—has to recover its excitability (Roger 2004). Interestingly, *anisotropic structures* of the cardiac muscle (such as *regional ischemia* or fibrotic regions) favour the discussed re-entrant *circular excitation* since anisotropy can lead to *re-entrant loops*. For *multiple re-entry mechanisms* in terms of *ventricular fibrillation* see Sect. 6.

It should be noted that the *premature ventricular contraction* may also induce *ventricular fibrillation* (Sect. 6), a life-threatening state. In particular, a *preceding* premature ventricular contraction exaggerates *non-uniformity in the recovery of excitability* in ventricles; compare with the *re-entry mechanism* from above. This non-uniformity elongates the *vulnerable period* (Sect. 6) and favours imminent fibrillation in response to another *proceeding* premature ventricular contraction during the vulnerable period. Interestingly, the vulnerable period can even extend beyond the T wave, i.e., beyond the repolarization phase of ventricles (Sect. 2.4.2).

The *premature ventricular contraction* is characterized by an *abnormally deformed, widened, and strong QRS complex* in the electrocardiogram while the P wave is usually missing (Sect. 2.4.2). The widening of the QRS complex is basically due to relatively slow propagation of the excitation through cardiac muscles as compared to (normal and fast) propagation along the *conductive system* in the heart (Fig. 2.35). The deformation of the QRS complex is caused by the different pathway of the excitation wave front in ventricles as compared to the normal pathway beginning at the atrioventricular node.

In contrast to premature ventricular contraction, *premature atrial contraction* is an abnormal heart beat initiated in atria by a pacemaker (i.e., prematurely depolarized region in an atrium) other than the sinoatrial node. The premature atrial contraction is characterized by an *abnormally shaped P wave* (and usually shortened PR interval) and the normal narrow QRS complex in the electrocardiogram. This is because this premature beat is initiated outside the sinoatrial node but the excitation propagates still normally through the atrioventricular node into ventricles (Sect. 2.4.2).

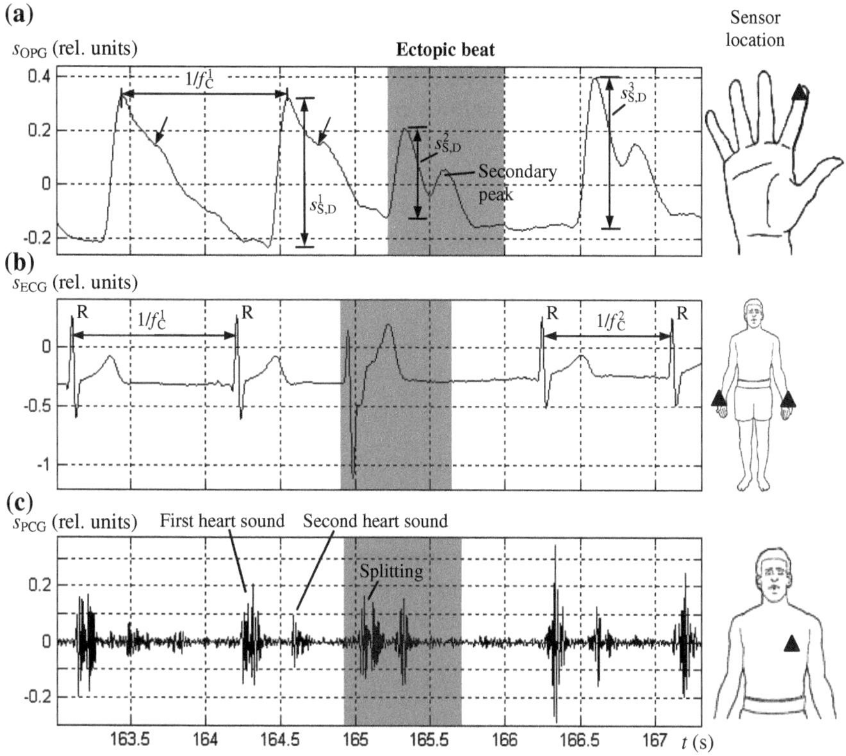

Fig. 5.32 Manifestation of an ectopic beat in different biosignals. (**a**) Optic biosignal optoplethysmogram s_{OPG} (from a fingertip on the right hand) with the optical sensor operated in the reflectance mode (Fig. 5.22b) and indicated instantaneous heart rate f_C. *Arrow* mark inflection points as approximate starts of the reflected waves (compare Fig. 2.48). (**b**) Electric biosignal electrocardiogram s_{ECG} (lead I Einthoven) with indicated R peaks. (**c**) Acoustic biosignal phonocardiogram s_{PCG} (from the heart region on the chest). *Grey background* indicates the appearance of the ectopic beat

normal heart beat. Usually, the subsequent heart beat tends to show increased $s_{S,D}$ ($s_{S,D}^3 \geq s_{S,D}^1$ in Fig. 5.32a), following a prolonged ventricular filling (Fig. 2.38).

The time period between the *ectopic beat* and the subsequent *normal heart beat* is elongated (> $1/f_C^1$, known as *compensatory pause*) in comparison to the normal interbeat interval before the ectopic beat (= $1/f_C^1$ in Fig. 5.32). Likewise, the interval between the preceding and proceeding normal heart beats—surrounding the premature ventricular contraction—stays almost the same as normal, namely, twice the normal interbeat interval (= $2/f_C^1$). Since the premature ventricular contraction arises before the normal heart beat was supposed to occur (Footnote 61), the *compensatory pause* is clearly longer than the *normal interbeat interval*.

In addition, a temporarily increased f_C immediately after the ectopic beat can be observed in Fig. 5.32 ($f_C^2 > f_C^1$). This is because the body compensates for the temporarily decreased stroke volume and blood flow (2.30) during the *ectopic beat*, the mechanism known as *heart rate turbulence* (Sect. 3.2.2.1). In terms of the electrocardiogram (Fig. 5.32b), a *widened QRS complex* can be observed during the ectopic beat, as described in Footnote 61. Furthermore, Fig. 5.32c shows that morphological properties of the *first heart sound* during the ectopic beat differ from those during the preceding and proceeding normal heart beats. In particular, the splitting of the first heart sound seems to dominate during the ectopic beat because the normal sequence of the valve's closure is disrupted by the ectopic beat; compare Fig. 3.32. Lastly, it should be recalled that there is an obvious time delay between the optoplethysmogram (Fig. 5.32a) and the electrocardiogram (Fig. 5.32b). This delay is a result of the relatively low pulse wave propagation velocity along vessels; for details see Fig. 3.2a, b and Sect. 3.1.1.

5.2.2.2 Respiratory Activity

Respiratory activity manifests clearly within the optic biosignal *optoplethysmogram*, as derived and discussed in section "Respiratory Activity" in Sect. 5.1.2.3. In fact, both the *pulsatile alternating component* I_{AC} and the *non-pulsatile direct component* I_{DC} of the *transmitted light intensity* I reflect respiratory activity. The corresponding *basic principle* is summarized in Fig. 5.17 with a typical optoplethysmogram illustrated in Fig. 5.15c. An (important) vital physiological parameter of respiratory activity is the respiratory rate f_R (Sect. 3.1.2).

Figure 5.33 demonstrates the *registration of* f_R out of the optoplethysmogram during normal breathing. In the *time domain* (Fig. 5.33b), the respiratory modulation of the pulsatile component of the optoplethysmogram can be observed. Here it should be recalled that the level of the optoplethysmogram is inversely related to the intensity I or, in other words, is proportional to the light absorption strength (Fig. 5.15c). In particular, the *systolic-diastolic deflection* $s_{S,D}$ (compare Fig. 5.15a) decreases *during inspiration*, as illustrated in Fig. 5.33d. It indicates reduced pulsation of the light absorption during inspiration; compare with the identical observation in Fig. 5.15b. During expiration, the reverse is true. For comparison, the *reference signal for inspiration and expiration* phases of the respiration cycle is depicted in Fig. 5.33d.

Furthermore, the *non-pulsatile component* of the optoplethysmogram (Fig. 5.33c) seems to behave in synchrony with the *systolic-diastolic deflection* (Fig. 5.33d). That is, the *non-pulsatile absorption* of light decreases *during inspiration*; likewise, the component I_{DC} increases. An identical behaviour of this non-pulsatile absorption was already observed in Fig. 5.15b, i.e., decreasing absorption with inspiration and increasing one with expiration.

The resulting oscillation rate of the non-pulsatile absorption (Fig. 5.33c) is actually the *respiratory rate* f_R which estimated time course is shown in Fig. 5.33e. For instance, signal processing methods in the time domain can be used to derive the instantaneous f_R; see Footnote 61 in Sect. 4. For the sake of completeness, it should be mentioned that the rate f_R can also be derived from the time course of $s_{S,D}$ (Fig. 5.33d) as well as from the previously derived time course of f_C (as exemplified in Fig. 5.31b).

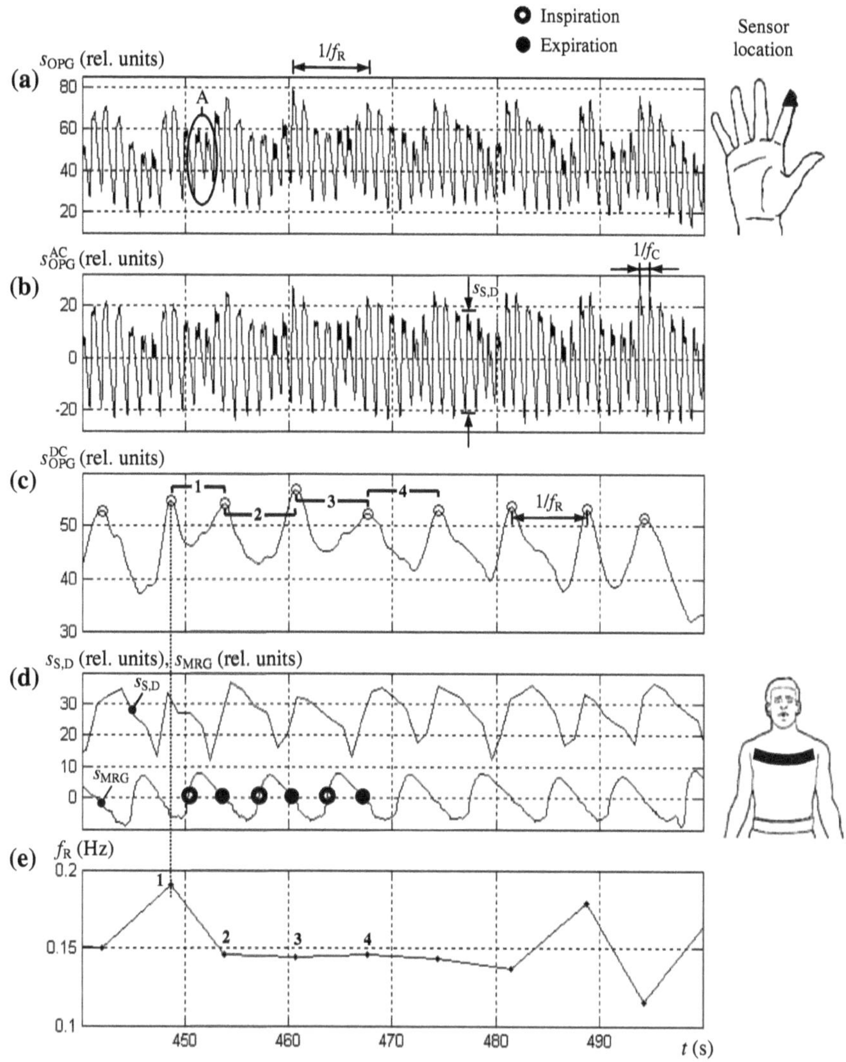

Fig. 5.33 Assessment of respiratory activity by the optic biosignal in the course of normal breathing. (**a**) Optic biosignal optoplethysmogram s_{OPG} (from a fingertip on the right hand) with the optical sensor operated in the reflectance mode (Fig. 5.22b). (**b**) The pulsatile (alternating) component s_{OPG}^{AC} of the total biosignal s_{OPG} (= $s_{OPG}^{AC} + s_{OPG}^{DC}$), which oscillates with the heart rate f_C. The component s_{OPG}^{AC} was isolated from s_{OPG} by a high-pass filter with the cut-off frequency of 0.25 Hz (Footnote 38). (**c**) The non-pulsatile (direct) component s_{OPG}^{DC} of s_{OPG}, isolated by a low-pass filter with the cut-off frequency of 0.25 Hz. (**d**) The systolic-diastolic deflection $s_{S,D}$ of s_{OPG}^{AC} (Fig. 5.15a) and the reference signal s_{MRG} for respiration which is given by mechanic biosignal mechanorespirogram (from chest circumference changes). (**e**) The instantaneous respiratory rate f_R derived from the time course of s_{OPG}^{DC} using signal processing methods in the time domain (Footnote 61 in Sect. 4). Four pairs of s_{OPG}^{DC} peaks and the corresponding values of f_R are denoted by *numbers*

In the latter case, the respiration-induced variations of f_C can be utilised in terms of dominant *cardiorespiratory interrelations* (Sect. 3.2.1).

Morphological changes in the *pulsatile waveform of the optoplethysmogram* reveal (temporary and permanent) changes in the *vessel stiffness* and *blood pressure* within the vessel. In accordance with (5.14), the deflection $s_{S,D}$ of the optoplethysmogram—or the temporary (pulsatile) change in the local light absorption—becomes less with decreasing *(pulsatile) volume change* of the transilluminated vessel (Fig. 5.14c). Likewise, the deflection amplitude of the optoplethysmogram becomes less with increasing *stiffness* of the vessel and decreasing change in *blood pressure* (during the cardiac cycle). Here it should be recalled that blood pressure is subject to change due to various physiological phenomena such as respiration-induced fluctuations of the stroke volume (Sect. 3.2).

In particular, the *waveform shape* of the optoplethysmogram discloses the location of the *reflected wave* (or even multiple waves) of blood pressure with respect to its incident wave (Fig. 2.53). The reflected waves can be recognized based on (possible) *secondary peaks or inflection points* in the waveform (Fig. 2.48). Consequently, numerous *short-term and long-term physiological phenomena* which impact the vessel stiffness and blood pressure leave their marks on the reflected wave, namely, on its amplitude and its time delay related to the onset of the incident wave. For instance, the *respiration impact* on reflections in the optoplethysmogram is shown in Figs. 3.31c, d and 3.36, whereas the *aging impact* is shown in Fig. 2.51. The dominant physiological phenomena are discussed in Sects. 2.5.2 and 3.2.1 in depth. Also Fig. 5.33 reveals this respiratory impact, in which double peaks within a single cardiac cycle become prominent in the optoplethysmogram during *inspiration* and (periodically) disappear during *expiration*; see double peaks within the region A in Fig. 5.33a. In other words, the *secondary peak* (right) after the primary systolic peak becomes prominent during inspiration.

Figure 5.32a illustrates a dominant secondary peak in the optoplethysmogram during the *ectopic beat*, whereas this peak is much less prominent in the neighbouring normal heart beats. As discussed in Sect. 5.2.2.1, there is a temporary drop in the *systolic-diastolic blood volume* (at the site of the sensor application) during the ectopic beat, which yields a temporary drop in the local *systolic-diastolic blood pressure*. Consequently, the propagation of the *reflected wave* (of blood pressure) is slowed down, in the course of which the incident wave overlaps with the reflected wave to a lesser extent than during the normal heart beat. This slowdown uncovers the *reflected wave as a secondary peak* in the waveform of the optoplethysmogram. That is,

- the *normal heart beat preceding the ectopic beat* shows only an *inflection point* after the systolic primary peak as the only indicator of the present (early) *reflected wave* (see arrows in Fig. 5.32a);
- the *ectopic beat* itself shows a clear (delayed) secondary peak as the *reflected wave* because the reflected wave is strongly delayed with respect to the incident wave; and
- the *normal heart beat proceeding the ectopic beat* shows again a less dominant (already less delayed) secondary peak as a signature of the *reflected wave*.

5.2.2.3 Blood Oxygenation

Blood oxygenation can be derived from the optic biosignal *optoplethysmogram*, as discussed in section "Blood Oxygenation" in Sect. 5.1.2.3. In fact, both the *pulsatile alternating component* I_{AC} and the *non-pulsatile direct component* I_{DC} of the *transmitted light intensity* I at (at least) two wavelengths of light are required for the estimation of the *hemoglobin oxygen saturation* S. The *basic principle* of this estimation is summarized in Fig. 5.19, whereas the physiological relevance of blood oxygenation is described in Sect. 3.1.4.

Figure 5.34 demonstrates the *registration of* S out of the optoplethysmogram in response to *voluntary breath holding*, i.e., *voluntary apnea*, compare Fig. 3.20. Two consecutive periods of breath holding are shown, one lasting for about 50 s and another for about 100 s. In the *time domain*, temporary *desaturations* can be observed (Fig. 5.34b), which occur with a significant *time delay* (see below) related to periods of breath holding (Fig. 5.34a). The latter periods of the effective breath holding are clearly depicted by the *reference signal for respiration* (Fig. 5.34a), whereas the reference is given by a mechanic biosignal recorded from the chest; compare Fig. 5.35.

In analogy with Figs. 5.33c, d, Fig. 5.34c illustrates the time course of the *non-pulsatile absorption of light* during (temporarily) interrupted breathing while Fig. 5.34d illustrates the corresponding time course of the *systolic-diastolic deflection* $s_{S,D}$. It can be observed that $s_{S,D}$ tends to decrease during apneas, as already observed in Fig. 3.37d. That is, the pulsatile blood volume in the periphery is reduced due to the vasoconstriction of peripheral arterial vessels during apneas; see section "Ceased respiration" in Sect. 3.2.1.1 and Footnote 239 in Sect. 3 for *diving reflex*. Likewise, *sympathetic activation* (of smooth muscles in the vessel's wall) is denoted by reduced $s_{S,D}$. During periods of normal breathing, the course of $s_{S,D}$ *fluctuates* with the respiratory rate f_R (Fig. 5.34d), as already observed in Fig. 5.33d, whereas this respiration-induced fluctuation obviously disappears when holding breath.

The behaviour of the derived *heart rate* f_C during breath holding is shown in Fig. 5.34e; compare with Fig. 5.31b. As discussed in section "Ceased respiration" in Sect. 3.2.1.1, this behaviour is less regular during apneic periods because the actual level of f_C is determined by the actual *balance between sympathetic and parasympathetic tone* (Fig. 3.37c). In short, arrested oxygen supply and progressively increasing urge to breath tend to increase *sympathetic tone* and thus to increase f_C. At the same time, the body tends to reduce the consumption of the remaining oxygen (especially in the heart muscles) by increasing *parasympathetic tone* and thus by decreasing f_C; compare Sect. 3.2.2.1. In the shown case of Fig. 5.34e, there is a slight tendency of f_C to increase during breath holding, especially during the first voluntary apnea. During periods of normal breathing, the instantaneous f_C *fluctuates* with the rate f_R, as already observed in Fig. 5.31b, whereas this respiration-induced fluctuation disappears when holding breath.

Figure 5.35 compares the *response time* of two optical sensors, whereas one sensor is applied on a *distal region* (namely, the finger, see Fig. 5.35a) and another

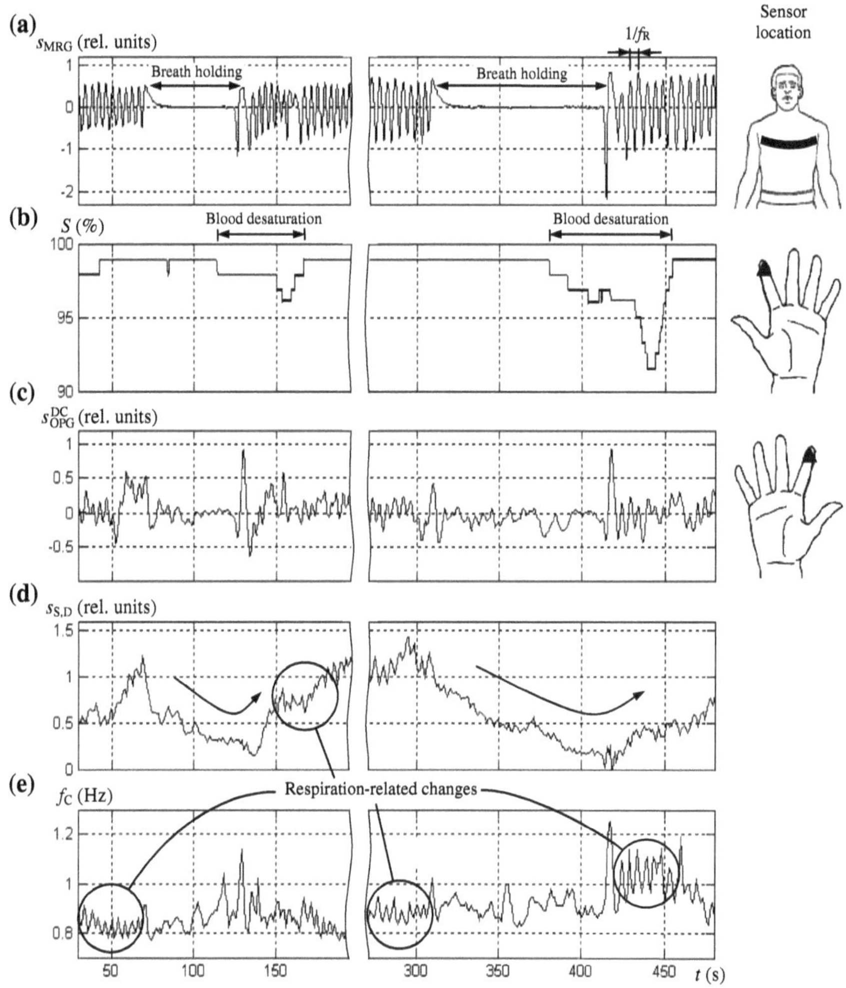

Fig. 5.34 Influence of voluntary breath holding on the optic biosignal. (**a**) Mechanic biosignal mechanorespirogram s_{MRG} (from chest circumference changes) as the reference signal for respiration with indicated respiratory rate f_R. (**b**) Temporal changes of hemoglobin oxygen saturation S, i.e., two consecutive and temporary desaturations, as derived from optic biosignals (from a finger on the left hand) with the optical sensor operated in the transmittance mode (Fig. 5.22a). (**c**) The non-pulsatile (direct) component s_{OPG}^{DC} of optic biosignal optoplethysmogram s_{OPG} with the optical sensor operated in the reflectance mode (from a fingertip on the right hand). The component s_{OPG}^{DC} was isolated by a low-pass filter with the cut-off frequency of 0.25 Hz (Footnote 38). (**d**) The systolic-diastolic deflection $s_{S,D}$ of the pulsatile (alternating) component s_{OPG}^{AC} of the total optic biosignal s_{OPG} (= s_{OPG}^{AC} + s_{OPG}^{DC}); compare Fig. 5.33b. The component s_{OPG}^{AC} was isolated from s_{OPG} by a high-pass filter with the cut-off frequency of 0.25 Hz. (**e**) The instantaneous heart rate f_C derived from the time course of s_{OPG} using signal processing methods in the time domain (Footnote 61 in Sect. 4)

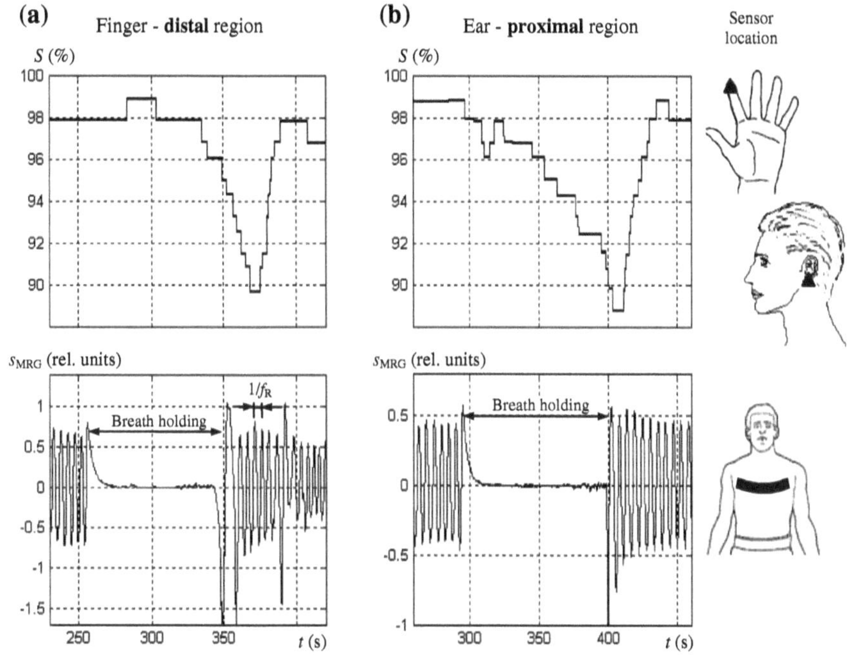

Fig. 5.35 Voluntary breath holding with the optical sensor applied on (**a**) a distal location of the body (i.e., the index finger on the left hand) and (**b**) a proximal location of the body (i.e., the left earlobe). *Upper subfigures* depict temporary changes of hemoglobin oxygen saturation S derived from the corresponding optic biosignals with the optical sensor operated in the transmittance mode (Fig. 5.22a). *Lower subfigures* depict mechanic biosignal mechanorespirogram s_{MRG} (from chest circumference changes) as the reference signal for respiration with indicated respiratory rate f_{R}

sensor is applied on a *proximal region* (namely, the earlobe, see Fig. 5.35b). That is, the *time delay* between the start of breath holding—as illustrated by the *reference signal for respiration* (lower subfigures in Fig. 5.35)—and the start of the estimated blood desaturation (upper subfigures) is obviously longer for the distal application region. In fact, the blood oxygenation in the *finger* starts to drop only after about 70 s after the breathing has ceased. In contrast, the blood oxygenation in the *earlobe* starts to drop almost instantaneously with the onset of the *voluntary apnea*. The observed qualitative differences in these time delays are in full agreement with the principles of the response time of optical oximeters (i.e., the response time as a function of the sensor application region), as laid down in section "Specific Issues" in Sect. 5.1.2.3 and "Application Regions" in Sect. 5.2.1.2 in depth.

Finally, it should be recalled that the estimation of *blood oxygenation* is subject to *diverse limitations* such as *motion artefacts*, *contacting force*, and poor *blood perfusion* (sections "Specific Issues" in Sect. 5.1.2.3 and "Motion Artefacts" in Sect. 5.1.2.3).

References

H.H. Asada, P. Shaltis, A. Reisner, R. Sokwoo, R.C. Hutchinson, Mobile monitoring with wearable photoplethysmographic biosensors. IEEE Eng. Med. Biol. Mag. **22**(3), 28–40 (2003)

A.A. Awad, M.A. Ghobashy, W. Ouda, R.G. Stout, D.G. Silverman, K.H. Shelley, Different responses of ear and finger pulse oximeter wave form to cold pressor test. Anesth. Analg. **92** (6), 1483–1486 (2001)

A.N. Bashkatov, E.A. Genina, V.I. Kochubey, V.V. Tuchin, Optical properties of human skin, subcutaneous and mucous tissues in the wavelength range from 400 to 2,000 nm. J. Phys. D Appl. Phys. **38**, 2543–2555 (2005)

D.E. Bebout, P.D. Mannheimer, C.C. Wun, Site-dependent differences in the time to detect changes in saturation during low perfusion. Crit. Care Med. **29**(12), 115 (2001)

D.A. Benaron, I.H. Parachikov, W.F. Cheong, S. Friedland, B.E. Rubinsky, D.M. Otten, F.W. Liu, C.J. Levinson, A.L. Murphy, J.W. Price, Y. Talmi, J.P. Weersing, J.L. Duckworth, U.B. Hörchner, E.L. Kermit, Design of a visible-light spectroscopy clinical tissue oximeter. J. Biomed. Opt. **10**(4), 44005-1–44005-9 (2005)

L. Bernardi, A. Radaelli, P.L. Solda, A.J. Coats, M. Reeder, A. Calciati, C.S. Garrard, P. Sleight, Autonomic control of skin microvessels: assessment by power spectrum of photoplethysmographic waves. Clin. Sci. **90**(5), 345–355 (1996)

L. Bernardi, M. Rossi, S. Leuzzi, E. Mevio, G. Fornasari, A. Calciati, C. Orlandi, P. Fratino, Reduction of 0.1 Hz microcirculatory fluctuations as evidence of sympathetic dysfunction in insulin-dependent diabetes. Cardiovasc. Res. **34**(1), 185–191 (1997)

Biomedical Optics Research Laboratory, University College London, Department of Medical Physics and Bioengineering, www.ucl.ac.uk (2010)

D. Chamier, *Unpublished ball pen drawing* (Institute of Art and Design, Vienna University of Technology, Vienna, 2014)

W.F. Cheong, S.A. Prahl, A.J. Welch, A review of the optical properties of biological tissues. IEEE J. Quantum Electron. **26**(12), 2166–2185 (1990)

C. Clark, R. Vinegar, J.D. Hardy, Goniometric spectrometer for the measurement of diffuse reflectance and transmittance of skin in the infrared spectral region. J. Opt. Soc. Am. **43**(11), 993–998 (1953)

G. Comtois, Y. Mendelson, P. Ramuka, A comparative evaluation of adaptive noise cancellation algorithms for minimizing motion artifacts in a forehead-mounted wearable pulse oximeter, in *Proceedings of the 29th Annual EMBS International Conference*, pp. 1528–1531 (2007)

M. Cope, The application of near infrared spectroscopy to non invasive monitoring of cerebral oxygenation in the newborn infant. Thesis at the University College London (1991)

R.J. Davies-Colley, W.N. Vant, D.G. Smith, *Colour and Clarity of Natural Waters* (The Blackburn Press, Caldwell, 2003)

German Institute for Standardization (in German: Deutsches Institut für Normung), norm DIN 33403-3 on climate at the workplace and its environments (2011)

A. Duncan, J.H. Meek, M. Clemence, C.E. Elwell, L. Tyszczuk, M. Cope, D.T. Delpy, Optical pathlength measurements on adult head, calf and forearm and the head of the newborn infant using phase resolved optical spectroscopy. Phys. Med. Biol. **40**(2), 295–304 (1995)

S. Duun, R.G. Haahr, K. Birkelund, P. Raahauge, P. Petersen, H. Dam, L. Noergaard, E.V. Thomsen: a novel ring shaped photodiode for reflectance pulse oximetry in wireless applications. IEEE Sens. **15**, 596–599 (2007)

C. Elwell, J. Hebden, Near-infrared spectroscopy. Biomed. Opt. Res. Lab. http://www.medphys.ucl.ac.uk/research/borl/research/NIR_topics/nirs.htm (1999)

Y.S. Fawzi, A.M. Youssef, M.H. El-Batanony, Y.M. Kadah, Determination of the optical properties of a two-layer tissue model by detecting photons migrating at progressively increasing depths. Appl. Opt. **42**(31), 6398–6411 (2003)

M.A. Franceschini, E. Gratton, D. Hueber, S. Fantini, Near-infrared absorption and scattering spectra of tissues in vivo. Proc. Soc. Opt. Eng. **3597**, 526–531 (1999a)

M.A. Franceschini, E. Gratton, S. Fantini, Noninvasive optical method of measuring tissue and arterial saturation: An application to absolute pulse oximetry of the brain. Opt. Lett. **24**(12), 829–831 (1999b)

M.A. Franceschini, D.A. Boas, A. Zourabian, S.G. Diamond, S. Nadgir, D.W. Lin, J.B. Moore, S. Fantini, Near-infrared spiroximetry: noninvasive measurements of venous saturation in piglets and human subjects. J. Appl. Physiol. **92**(1), 372–384 (2002)

C. Furse, D.A. Christensen, C.H. Durney, *Basic Introduction to Bioelectromagnetics* (CRC Press Publisher, Boca Raton, 2009)

E.A. Genina, A.N. Bashkatov, V.V. Tuchin, Optical clearing of cranial bone. Adv. Opt. Technol. ID **267867**, 1–8 (2008)

D.C. Giancoli, *Physics* (In German: Physik) (Pearson Studium Publisher, Munich, 2006)

J.M. Goldman, M.T. Petterson, R.J. Kopotic, S.J. Barker, Masimo signal extraction pulse oximetry. J. Clin. Monit. Comput. **16**(7), 475–483 (2000)

J.D. Hardy, H.T. Hammel, D. Murgatroyd, Spectral transmittance and reflectance of excised human skin. J. Appl. Physiol. **9**(2), 257–264 (1956)

E. Hill, M.D. Stoneham, Practical applications of pulse oximetry. World Anaesth. Online **11**(4), 1–2 (2000)

E.M.C. Hillman, Experimental and theoretical investigations of near infrared tomographic imaging methods and clinical applications. Thesis at the University of London (2002)

W. Hoppe, W. Lohmann, H. Markl, H. Ziegler, *Biophysik* (Springer, Berlin, 1982)

HyperPhysics Project, http://hyperphysics.phy-astr.gsu.edu/hbase/hph.html (2012)

International Commission on Non-Ionizing Radiation Protection, ICNIRP guidelines on limits of exposure to broad-band incoherent optical radiation (0.38–3 μm). Health Phys. **73**(3), 539–554 (1997)

International Commission on Non-Ionizing Radiation Protection, ICNIRP statement on far infrared radiation exposure. Health Phys. **91**(6), 630–645 (2006)

S.L. Jacques, Skin optics. Online Oregon Medical Laser Center http://omlc.ogi.edu/news/jan98/skinoptics.html (1998)

S.L. Jacques, S.A. Prahl, Introduction to biomedical optics. Oregon Graduate Institute http://omlc.bme.ogi.edu/classroom/ece532/index.html (2002)

S.L. Jacques, *Optical Properties of Tissue*. Tutorial Lecture at the Graduate Summer School Bio-Photonics in Sweden (2003)

A. Jubran, Pulse oximetry. Crit. Care **3**(2), R11–R17 (1999)

V. Kamat, Pulse oximetry. Indian J. Anaesth. **46**(4), 261–268 (2002)

E. Kaniusas, *Multiparametric Physiological Sensors*. Habilitation theses at the Vienna University of Technology (2006)

E. Kaniusas, H. Pfützner, L. Mehnen, Optical tissue absorption sensor on the thorax: Possibilities and restrictions. Int. J. Appl. Electromagn. Mech. **25**(1–4), 649–655 (2007)

E. Kaniusas, G. Varoneckas, B. Mahr, J.C. Szeles, Optic visualisation of auricular nerves and blood vessels: optimisation and validation. IEEE Trans. Instrum. Meas. **60**(10), 3253–3258 (2011)

J.M. Kim, K. Arakawa, K.T. Benson, D.K. Fox, Pulse oximetry and circulatory kinetics associated with pulse volume amplitude measured by photoelectric plethysmography. Anesth. Analg. **65** (12), 1333–1339 (1986)

V. König, R. Huch, A. Huch, Reflectance pulse oximetry—principles and obstetric application in the Zurich system. J. Clin. Monit. Comput. **14**, 403–412 (1998)

H. Krieger, *Basics of Radiation Physics and Radiation Protection* (In German: Grundlagen der Strahlungsphysik und des Strahlenschutzes) (Teubner Publisher, Leipzig, 2004)

R. Krishnan, B. Natarajan, S. Warren, Two-stage approach for detection and reduction of motion artifacts in photoplethysmographic data. IEEE Trans. Biomed. Eng. **57**(8), 1867–1876 (2010)

G. Kumar, J.M. Schmitt, Optimal probe geometry for near infrared spectroscopy of biological tissue. Appl. Opt. **36**(10), 2286–2293 (1997)

P.A. Kyriacou, S. Powell, R.M. Langford, D.P. Jones, Esophageal pulse oximetry utilizing reflectance photoplethysmography. IEEE Trans. Biomed. Eng. **49**(11), 1360–1368 (2002)

J.C. Lai, Y.Y. Zhang, Z.H. Li, H.J. Jiang, A.Z. He, Complex refractive index measurement of biological tissues by attenuated total reflection ellipsometry. Appl. Opt. **49**(16), 3235–3238 (2010)

J. Ling, S. Takatani, G.P. Noon, Y. Nose, In-vivo studies of reflectance pulse oximeter sensor. Proc. Int. Soc. Opt. Photonics (SPIE) **1887**, 256–262 (1993)

P.D. Mannheimer, J.R. Casciani, M.E. Fein, S.L. Nierlich, Wavelength selection for low-saturation pulse oximetry. IEEE Trans. Biomed. Eng. **44**(3), 148–158 (1997)

B. Manzke, J. Schwider, N. Lutter, K. Engelhardt, W. Stork, Multiwavelength pulse oximetry in the measurement of hemoglobin fractions. Proc. Soc. Opt. Eng. **2676**, 332–340 (1996)

S.A. Mascaro, H.H. Asada, Photoplethysmograph fingernail sensors for measuring finger forces without haptic obstruction. IEEE Trans. Robot. Autom. **17**(5), 698–708 (2001)

I.V. Meglinski, S.J. Matcher, Quantitative assessment of skin layers absorption and skin reflectance spectra simulation in the visible and near-infrared spectral regions. Physiol. Meas. **23**(4), 741–753 (2002)

Y. Mendelson, B.D. Ochs, Noninvasive pulse oximetry utilizing skin reflectance photoplethysmography. IEEE Trans. Biomed. Eng. **35**(10), 798–805 (1988)

P.M. Middleton, C.H.H. Tang, G.S.H. Chan, S. Bishop, A.V. Savkin, N.H. Lovell, Peripheral photoplethysmography variability analysis of sepsis patients. Med. Biol. Eng. Compu. **49**(3), 337–347 (2011)

J.R. Mourant, J.P. Freyer, A.H. Hielscher, A.A. Eick, D. Shen, T.M. Johnson, Mechanisms of light scattering from biological cells relevant to noninvasive optical-tissue diagnosis. Appl. Opt. **37** (16), 3586–3593 (1998)

Specific extinction spectra of tissue chromophores, Online Biomedical Optics Research Laboratory http://www.medphys.ucl.ac.uk/research/borl/research/NIR_topics/spectra/spectra.htm (2003)

H. Pfützner, *Applied Biophysics* (In German: Angewandte Biophysik) (Springer, Berlin, 2003)

S. Prahl, Optical absorption of hemoglobin. Online Oregon Medical Laser Center http://omlc.bme.ogi.edu/spectra/hemoglobin/index.html (1999)

S. Reichelt, J. Fiala, A. Werber, K. Forster, C. Heilmann, R. Klemm, H. Zappe, Development of an implantable pulse oximeter. IEEE Trans. Biomed. Eng. **55**(2), 581–588 (2008)

G.M. Roger, *Clinical Electrocardiography and Arrhythmias*. Lecture notes from the Massachusetts Institute of Technology (2004)

D.K. Sardar, L.B. Levy, Optical properties of whole blood. Lasers Med. Sci. **13**(2), 106–111 (1998)

J.M. Schmitt, Simple photon diffusion analysis of the effects of multiple scattering on pulse oximetry. IEEE Trans. Biomed. Eng. **38**(12), 1194–1203 (1991)

D.J. Segelstein, The complex refractive index of water. Theses at the University of Missouri-Kansas City (1981)

M. Shafique, P.A. Kyriacou, S.K. Pal, Investigation of photoplethysmographic signals and blood oxygen saturation values on healthy volunteers during cuff-induced hypoperfusion using a multimode PPG/SpO2 sensor. Med. Biol. Eng. Compu. **50**, 575–583 (2012)

C.R. Simpson, M. Kohl, M. Essenpreis, M. Cope, Near-infrared optical properties of ex vivo human skin and subcutaneous tissues measured using the monte carlo inversion technique. Phys. Med. Biol. **43**, 2465–2478 (1998a)

C.R. Simpson, J. Laufer, M. Kohl, M. Essenpreis, M. Cope, Measurement of skin optical properties. Online Biomedical Optics Research Laboratory http://www.medphys.ucl.ac.uk/research/borl/research/NIR_topics/skin/skinoptprop.htm (1998b)

J. Sola, O. Chetelat, J. Krauss, On the reliability of pulse oximetry at the sternum, in *Proceedings of the 29th Annual EMBS International Conference*, pp. 1537 (2007a)

J. Sola, O. Chetelat, Combination of multiple light paths in pulse oximetry: the finger ring example. in *Proceedings of the 29th Annual EMBS International Conference*, pp. 6697–6698 (2007b)

J. Sola, S. Castoldi, O. Chetelat, M. Correvon, S. Dasen, S. Droz, N. Jacob, R. Kormann, V. Neumann, A. Perrenoud, P. Pilloud, C. Verjus, G. Viardot, SpO2 sensor embedded in a finger ring: design and implementation, in *Proceedings of the 28th Annual EMBS International Conference*, pp. 4295–4298 (2006)

R. Splinter, B.A. Hooper, *An Introduction to Biomedical Optics* (Publisher Taylor and Francis, Oxford, 2007)

S. Takatani, J. Ling, Optical oximetry sensors for whole blood and tissue. IEEE Eng. Med. Biol. Mag. **13**(3), 347–357 (1994)

X.F. Teng, Y.T. Zhang, The effect of contacting force on photoplethysmographic signals. Physiol. Meas. **25**(5), 1323–1335 (2004)

T. Torfs, V. Leonov, R.J.M. Vullers, Pulse oximeter fully powered by human body heat. Sens. Transducers J. **80**(6), 1230–1238 (2007)

V.V. Tuchin, Optical clearing of tissues and blood using the immersion method. J. Phys. D Appl. Phys. **38**(15), 2497–2518 (2005)

N. Ugryumova, S.J. Matcher, D.P. Attenburrow, Measurement of bone mineral density via light scattering. Phys. Med. Biol. **49**, 469–483 (2004)

B. Venema, N. Blanik, V. Blazek, H. Gehring, A. Opp, S. Leonhardt, Advances in reflective oxygen saturation monitoring with a novel in-ear sensor system: results of a human hypoxia study. IEEE Trans. Biomed. Eng. **59**(7), 2003–2010 (2012)

C.Y. Wang, M.L. Chuang, S.J. Liang, J. Tsai, C.C. Chuang, Y.S. Hsieh, C.W. Lu, P.L. Lee, C.W. Sun, Diffuse optical multipatch technique for tissue oxygenation monitoring: clinical study in intensive care unit. IEEE Trans. Biomed. Eng. **59**(1), 87–94 (2012)

Scattering by pure water. WET Labs http://www.wetlabs.com/iopdescript/purescatter2.htm (2005)

B. Winey, Y. Yan, Oximetry considerations in the small source detector separation limit, in *Proceedings of the 28th Annual EMBS International Conference*, pp. 1941–1943 (2006)

R.A. Winter, Arterial blood flow simulator. U.S. Patent Nr. 6400973 (2002)

M.W. Wukitsch, M.T. Petterson, D.R. Tobler, J.A. Pologe, Pulse oximetry: analysis of theory, technology, and practice. J. Clin. Monit. **4**(4), 290–301 (1988)

Y. Xu, N. Iftimia, H. Jiang, L.L. Key, M.B. Bolster, Imaging of in vitro and in vivo bones and joints with continuous-wave diffuse optical tomography. Opt. Express **8**(7), 447–451 (2001)

J.I. Youn, S.A. Telenkov, E. Kim, N.C. Bhavaraju, B.J.F. Wong, J.W. Valvano, T.E. Milner, Optical and thermal properties of nasal septal cartilage. Lasers Surg. Med. **27**(2), 119–128 (2000)

G. Zonios, U. Shankar, V.K. Iyer, Pulse oximetry theory and calibration for low saturations. IEEE Trans. Biomed. Eng. **51**(5), 818–822 (2004)

A. Zourabian, A. Siegel, B. Chance, N. Ramanujan, M. Rode, D.A. Boas, Trans-abdominal monitoring of fetal arterial blood oxygenation using pulse oximetry. J. Biomed. Opt. **5**(4), 391–405 (2000)

Index

E. Kaniusas, *Biomedical Signals and Sensors II*,
Biological and Medical Physics, Biomedical Engineering,
DOI 10.1007/978-3-662-45106-9